The Fragments of the Work of Heraclitus of Ephesus on Nature;

THE FRAGMENTS

OF THE WORK OF

HERACLITUS OF EPHESUS

ON NATURE

TRANSLATED FROM THE GREEK TEXT OF BYWATER,
WITH AN INTRODUCTION HISTORICAL AND CRITICAL

BY

G. T. W. PATRICK, Ph.D.

PROFESSOR OF PHILOSOPHY IN THE STATE UNIVERSITY OF IOWA

BALTIMORE
N. MURRAY
1889

THE FRAGMENTS

OF THE WORK OF

HERACLITUS OF EPHESUS

ON NATURE

TRANSLATED FROM THE GREEK TEXT OF BYWATER,
WITH AN INTRODUCTION HISTORICAL AND CRITICAL

BY

G. T. W. PATRICK, Ph.D.
PROFESSOR OF PHILOSOPHY IN THE STATE UNIVERSITY OF IOWA

BALTIMORE
N. MURRAY
1889

[Reprinted from the AMERICAN JOURNAL OF PSYCHOLOGY, 1888.]

A THESIS ACCEPTED FOR THE DEGREE OF DOCTOR OF PHILOSOPHY IN THE
JOHNS HOPKINS UNIVERSITY, 1888.

PRESS OF ISAAC FRIEDENWALD,
BALTIMORE.

Οἱ ῥέοντες.

I.

All thoughts, all creeds, all dreams are true,
    All visions wild and strange,
Man is the measure of all truth
    Unto himself. All truth is change,
All men do walk in sleep, and all
    Have faith in that they dream
For all things are as they seem to all,
    And all things flow like a stream.

II

There is no rest, no calm, no pause,
    Nor good nor ill, nor light nor shade,
Nor essence nor eternal laws:
    For nothing is, but all is made.
But if I dream that all these are,
    They are to me for that I dream,
For all things are as they seem to all,
    And all things flow like a stream.

    Argal—this very opinion is only true
relatively to the flowing philosophers.

TENNYSON.

PREFACE.

The latest writers on Heraclitus, namely, Gustav Teichmüller and Edmund Pfleiderer, have thought it necessary to preface their works with an apology for adding other monographs to the Heraclitic literature, already enriched by treatises from such distinguished men as Schleiermacher, Lassalle, Zeller, and Schuster. That still other study of Heraclitus, however, needs no apology, will be admitted when it is seen that these scholarly critics, instead of determining the place of Heraclitus in the history of philosophy, have so far disagreed, that while Schuster makes him out to be a sensationalist and empiricist, Lassalle finds that he is a rationalist and idealist. While to Teichmüller, his starting point and the key to his whole system is found in his physics, to Zeller it is found in his metaphysics, and to Pfleiderer in his religion. Heraclitus' theology was derived, according to Teichmüller, from Egypt; according to Lassalle, from India; according to Pfleiderer, from the Greek Mysteries. The Heraclitic flux, according to Pfleiderer, was consequent on his abstract theories; according to Teichmüller, his abstract theories resulted from his observation of the flux. Pfleiderer says that Heraclitus was an optimist; Gottlob

Mayer says that he was a pessimist. According to Schuster he was a hylozoist, according to Zeller a pantheist, according to Pfleiderer a panzoist, according to Lassalle a panlogist. Naturally, therefore, in the hands of these critics, with their various theories to support, the remains of Heraclitus' work have suffered a violence of interpretation only partially excused by his known obscurity. No small proportion of the fragments, as will be seen in my introduction, have been taken in a diametrically opposite sense.

Recently a contribution towards the disentanglement of this maze has been made by Mr. Bywater, an acute English scholar. His work (Heracliti Ephesii Reliquiae, Oxford, 1877) is simply a complete edition of the now existing fragments of Heraclitus' work, together with the sources from which they are drawn, with so much of the context as to make them intelligible.

Under these circumstances I have thought that a translation of the fragments into English, that every man may read and judge for himself, would be the best contribution that could be made. The increasing interest in early Greek philosophy, and particularly in Heraclitus, who is the one Greek thinker most in accord with the thought of our century, makes such a translation justifiable, and the excellent and timely edition of the Greek text by Mr. Bywater makes it practicable.

The translations both of the fragments and of the context are made from the original sources, though I have followed the text of Bywater except in a very few cases, designated in the critical notes. As a number of the fragments are ambiguous, and several of them contain a play upon words, I have appended the entire Greek text.

The collection of sources is wholly that of Mr. Bywater. In these I have made a translation, not of all the references, but only of those from which the fragment is immediately taken, adding others only in cases of especial interest.

My acknowledgments are due to Dr. Basil L. Gildersleeve, of the Johns Hopkins University, for kind suggestions concerning the translation, and to Dr. G. Stanley Hall for valuable assistance in relation to the plan of the work.

BALTIMORE, SEPTEMBER 1, 1888.

CONTENTS.

INTRODUCTION.

Section I.—Historical and Critical.

	PAGE
Literature	1
Over-systemization in Philosophy	2
Over-interpretation in Historical Criticism	3
Exposition of Lassalle	4
Hegel's Conception of Heraclitus	5
Criticism of Hegel's Conception	6
Criticism of Lassalle	9
Exposition of Schuster	11
Criticism of Schuster	17
Exposition of Teichmuller	23
Criticism of Teichmuller	31
Exposition of Pfleiderer	39
Criticism of Pfleiderer	46

Section II.—Reconstructive.

I.

Can the Positions of the Critics be harmonized?	56
Heraclitus' Starting-point	57
Heraclitus as a Preacher and Prophet	57
The Content of his Message	58
The Universal Order	60
Strife	62
The Unity of Opposites	63
The Flux	65
Cosmogony	68
Ethics	69
Optimism	71

II.

Cause of the Present Interest in Heraclitus 72
Passion for Origins 72
Greek Objectivity 73
Heraclitic Ideas 74
Relation to Socrates and Plato 75
Socrates 76
Birth of Self-consciousness 77
Loss of Love of Beauty 78
Rise of Transcendentalism 79
Platonic Dualism 80
Return to Heraclitus 82
Defeat of Heraclitus 83

TRANSLATION OF THE FRAGMENTS 84–114
CRITICAL NOTES 115–123
GREEK TEXT 124–131

INTRODUCTION.

Section I.—Historical and Critical.

Modern Heraclitic literature belongs wholly to the present century. The most important works are the following :—Schleiermacher : Herakleitòs, der Dunkle von Ephesos, in Wolf and Buttmann's Museum der Alterthumswissenschaft, Vol. I, 1807, pp. 313–533, and in Schleiermacher's Sämmt. Werke, Abth. III, Vol. 2, Berlin, 1838, pp. 1–146 ;—Jak. Bernays : Heraclitea, Bonn, 1848 ; Heraklitische Studien, in the Rhein. Mus., new series, VII, pp. 90–116, 1850 ; Neue Bruchstücke des Heraklit, ibid. IX, pp. 241–269, 1854 ; Die Heraklitischen Briefe, Berlin, 1869 ;—Ferd. Lassalle : Die Philosophie Herakleitos des Dunkeln von Ephesos, 2 vols., Berlin, 1858 ;—Paul Schuster : Heraklit von Ephesus, in Actis soc. phil. Lips. ed. Fr. Ritschelius, 1873, III, 1–397 ;—Teichmüller, Neue Stud. z. Gesch. der Begriffe, Heft I, Gotha, 1876, and II, 1878 ;—Bywater : Heracliti Ephesii Reliquiae, Oxford, 1877 ;—Edmund Pfleiderer : Die Philosophie des Heraklit von Ephesus im Lichte der Mysterienidee, Berlin, 1886 ;—Eduard Zeller : Die Philosophie der Griechen, Bd. I, pp. 566–677.

There may be mentioned also the following additional writings which have been consulted in the preparation of these pages :—Gottlob Mayer : Heraklit von Ephesus und Arthur Schopenhauer, Heidelberg, 1886 ; Campbell : Theaetetus of Plato, Appendix A, Oxford, 1883 ; A. W. Benn : The Greek Philosophers, London, 1882.

After the introductory collection and arrangement of the Heraclitic fragments by Schleiermacher, and the scholarly discriminative work and additions of Bernays, four attempts have been made successively by Lassalle, Schuster, Teichmüller, and Pfleiderer, to reconstruct or interpret the philosophical system of Heraclitus. The positions taken and the results arrived at by these eminent scholars and critics are largely, if not wholly, different and discordant. A brief statement of their several positions will be our best introduction to the study of Heraclitus at first hand, and at the same time will offer us incidentally some striking examples of prevalent methods of historic criticism.

One of the greatest evils in circles of philosophical and religious thought has always been the evil of *over-systemization*. It is classification, or the scientific method, carried too far. It is the tendency to arrange under any outlined system or theory, more facts than it will properly include. It is the temptation, in a word, to classify according to resemblances which are too faint or fanciful. In the field of historic criticism this evil takes the form of *over-interpretation*. Just as in daily life we interpret every sense perception according to our own mental forms, so we tend to read our own thoughts into every saying of the ancients, and then proceed to use these, often without dishonesty, to support our favorite modern systems. The use of sacred writings will naturally occur to every one as the most striking illustration of this over-interpretation. Especially in the exegesis of the Bible has this prostitution of ancient writings to every man's religious views been long since recognized and condemned, and if most recently this tendency has been largely corrected in religious circles, it is all the more deplorable, in philosophical criticism, to find it still flourishing. Unfortunately, this vice continues, and it appears nowhere more plainly than in the interpretation of Greek philosophy. There is a great temptation to. modern writers to use the Greek philosophers as props to support their own systems—a temptation to interpret them arbitrarily, to look down upon them patronizingly, as it were, showing that what they meant was this or that modern thought, having only not learned to express themselves as well as we have. Among historians of philosophy this appears as a one-sidedness, so that it is commonly necessary in reading a history of philosophy to make a correction for the author's "personal equation." The histories of Schwegler and of Lewes are examples—the one biased by Hegelianism, the other by Positivism. Undoubtedly, a certain personal equation is unavoidable, and it is as impossible for an interpreter of Greek philosophy to make himself wholly Greek as it is unfair to represent the ancient thinker as wholly German or English. But when this becomes complete one-sidedness, or blindness to all but one series of an author's thoughts, or a willful or even unintentional perversion of his words, vigorous remonstrance is called for.

This attempt to fully understand the ancients, to make them speak in the phraseology of some modern school, must be distinguished from the recent movement, represented by Prof. Lagarde and others, in interpreting historic thought and historic events psychologically. This movement is certainly legitimate, based as it is on the truth of the similarity of constitution of all human minds, and the probability that underlying all representative historic creeds are great related if not identical thoughts. Even here, of course, the attempt to express these thoughts in the set phrases of any one people is inadequate.

We proceed, then, to look at some of the work done upon the philosophy of Heraclitus. Here we shall not attempt any examination of Zeller's exposition, since his work, though it is perhaps the very best that has been done in this field, is critical rather than reconstructive, and like his whole history of Greek philosophy, is a marvel of candor as well as of immense research. Even Zeller, however, has not wholly escaped the charge of one-sidedness, since Benn, in the preface to his work on the Greek philosophers, has accused him of never having outgrown the semi-Hegelian prejudice of his youth.

LASSALLE.

Lassalle, in two ponderous volumes noted above (page 1), made the first and most elaborate attempt to reconstruct the system of the Ephesian philosopher. His work exhibits immense labor and study, and extended research in the discovery of new fragments and of ancient testimony, together with some acuteness in their use. Lassalle has a very distinct view of the philosophy of Heraclitus. But it is not an original view. It is, in fact, nothing but an expansion of the short account of Heraclitus in Hegel's History of Philosophy, although Lassalle makes no mention of him, except to quote upon his title-page Hegel's well-known motto, "Es ist kein Satz des Heraklit, den ich nicht in meine Logik aufgenommen." Hegel's conception of Heraclitus is, in a word, as follows: Heraclitus' Absolute was the unity of being and non-being. His whole system was an expansion of the speculative thought of the principle of pure becoming. He apprehended, and was the first to apprehend, the Absolute as a *process*, as the unity of opposites, as dialectic itself. His great contribution was the speculative transition from the being of the Eleatics to the idea of becoming. Now how does Hegel support this position? There is in his Logic but one passage referring to Heraclitus. There he says, "Glancing at the principle of the Eleatics, Heraclitus then goes on to say, 'Being no more is than non-being' (οὐδὲν μᾶλλον τὸ ὂν τοῦ μὴ ὄντος ἐστὶ), a statement expressing the negative nature of abstract being and its identity with non-being" (Wallace, The Logic of Hegel, p. 144; cp. Science of Logic, Hegel's Werke, Vol. 3, p. 80). Hegel omits, in the Logic, to give the reference to the above quotation, but in his History of Philosophy (Werke, Vol. 13, p. 332) he quotes the same passage with the reference. It is to Aristotle, Metaph. i. 4. We turn to the same and find that it is a passage which Aristotle quotes from the Atomists, Democritus and Leucippus, and that it has not the slightest reference to Heraclitus, who, indeed, is not mentioned in the same chapter. This is rather discouraging, but the account in the History of Philosophy, to which we now turn, is scarcely less so. There Hegel begins his exposition of Heraclitus as follows:

"1. Das allgemeine Princip. Dieser kühne Geist (Heraclitus) hat zuerst das tiefe Wort gesagt, 'Das Seyn ist nicht mehr als das Nichtseyn,' es ist ebenso wenig, oder, 'Seyn und Nichts sey dasselbe,' das Wesen sey die Veränderung" (Gesch. d. Phil. Vol. 13, p. 332).

Now it happens that Heraclitus said nothing of the kind. As references Hegel gives Aristotle, Metaphys. i. 4; iv. 7; iv. 3. The first passage, as we have already seen, is from the Atomists. The second turns out upon examination to be simply the expression,

"All things are and are not" (πάντα εἶναι καὶ μὴ εἶναι), and the third is a statement of Aristotle that some people supposed Heraclitus to have said that the same thing could both be and not be the same. Moreover, neither of these passages is Heraclitic in form, and they are not even mentioned in Bywater's edition. The only expression of Heraclitus that resembles in form the above passage from Aristotle is that of frag. 81, "Into the same river we step and we do not step. We are and we are not." The over-interpretation by which this simple passage, expressing incessant physical *change*, is transformed into the logical principle of Hegel, "Das Seyn ist nicht mehr als das Nichtseyn," "Seyn und Nichts sey dasselbe," is audacious at least. Furthermore, we may say here in passing, that neither the expressions τὸ ὄν, μὴ ὄν, nor even τὸ γιγνόμενον, occur in any genuine saying of Heraclitus; although if they did occur, it would be easy to show that they could not mean at all what Hegel meant by being, non-being, and becoming. Even the Eleatic Being was not at all the same with that of Hegel, but was finite, spherical, and something very much like that which we should call material. But Heraclitus, who indeed preceded Parmenides, said nothing of being nor of non-being, nor did he speak of becoming in the abstract, although the trustful reader of Hegel, Lassalle, or Ferrier, might well suppose he spoke of nothing else. That which these writers mistook for becoming was, as we shall see later, only physical *change*. With the loss of this corner-stone, the Heraclitic support of the Hegelian Logic fails, and Hegel's boast that there was no sentence of Heraclitus that his Logic had not taken up becomes rather ludicrous, especially if one will read through the remains of

Heraclitus' work on Nature and search for his rich and varied thoughts in the Logic of Hegel.

Returning now to Lassalle, the above principles are carried out more in detail as follows: The chief point in the philosophy of Heraclitus is that here first the formal notion of the speculative idea in general was grasped. With him first emerged the conception of pure thought defecated of the sensuous. His ground principle was the dialectical opposition of being and non-being. The kernel and whole depth of his philosophy may be expressed in the one sentence, "Only non-being is" (Lassalle, Vol. 1, p. 35). The unity of being and non-being is a unity of process (processirende Einheit). It is the unity of opposites, the idea of becoming, the divine law, the γνώμη of the determining God (Id. Vol. 1, p. 24). Fire, strife, peace, time, necessity, harmony, the way up and down, the flux, justice, fate, Logos, are all different terms for this one idea (Id. Vol. 1, p. 57). Hence arises Heraclitus' obscurity. It is not a mere grammatical obscurity, as Schleiermacher, following Aristotle (Rhet. iii. 5, p. 1407, b. 14) thought; nor is it a willful obscurity, but it arises from the very nature of his great thought, which could not be enunciated in exact terms, but could only be suggested by such words as fire, time, etc., and so he labored on with one new symbol after another, vainly trying to express himself.

The Heraclitic fire is a "metaphysical abstraction" —a pure process, "whose existence is pure self-annulling (sich aufheben), whose being is pure self-consumption (sich selbst verzehren)" (Lassalle, Vol. 1, p. 18).

Most clearly, however, is the great thought of Heraclitus shown in "the way up and down," which does not involve change of place, but only a logical process.

It is "nothing else" than the change from being into non-being and the reverse. The way down is transition into being; the way up is the return into the pure and free negativity of non-being, motion in the undisturbed ideal harmony (Id. Vol. 2, p. 241 ff.).

God, in his adequate form, is "nothing else" than pure negativity, the pure unity of process of opposites. Nature is only the corporeal manifestation of the law of the identity of opposites. It owes its existence to privation ($ἀδικα$), that is, to the injustice which pure becoming suffers when it becomes being (Id. Vol. 1, p. 138).

The $ἀναθυμίασις$ of Heraclitus is not any vapor or sensible exhalation, but is "nothing else" than the way up, or the $ἐκπύρωσις$, that is, the cessation of the sensible and the particular and the assumption of the real universal becoming. $Ἀναθυμιωμέναι$, Lassalle says, should be translated "processirend" (Id. Vol. 1, p. 144).

The Heraclitic flux is the same as the way up and down. It is the dialectic of spacial being; it is the unity of being and non-being as spacial; it is the here which is not here. The $περιέχον$ of Heraclitus is not anything physical or spacial, but "the universal real process of becoming," which works through the Logos or law of thought (Id. Vol. 1, p. 306).

The Heraclitic Logos is the pure intelligible logical law of the identity in process (die processirende Identität) of being and non-being. It is "nothing else" than the law of opposites and the change into the same (Id. Vol. 1, p. 327; Vol. 2, p. 265).

The substance of the soul is identical with the substance of nature. It is pure becoming which has incorporated itself, embraced the way down. The dry or fiery soul is better than the moist because moisture is "nothing else" than a symbol of the downward way. The soul that is moist has descended out of its pure self-annulling movement or negativity in process, into the sphere of the particular and determinate (Id. Vol. 1, pp. 180, 192).

Heraclitus, in his desperate labor to express this idea, enters the sphere of religion. Dionysus and Hades are the same, he says (see frag. 127). That is, says Lassalle, Dionysus, the god of generation which represents the descent of pure non-being into being, is identical with Hades, the god of death; and this fragment, which is a polemic against Dionysus, is really a polemic against being, which is inferior to non-being (Id. Vol. 1, p. 208).

Knowledge consists in the recognition in each particular thing of the two opposites which constitute its nature (Id. Vol. 2, p. 272). Of ethics, the formal principle is self-realization or self-representation. It is the realization of what we are in ourselves or according to our inner nature. The ideal is separation from the sensible and particular and the realization of the universal (Id. Vol. 2, p. 428 ff.).

Such in brief outline is what Ferdinand Lassalle finds in Heraclitus' book On Nature. As an exposition of Heraclitus it is not worth the space we have given it, or any space, in fact; but as one of the most beautiful illustrations of over-systemization, it is extremely valuable. Any formal refutation of his conception of Heraclitus is unnecessary, for almost the whole of it is without any foundation whatever. The expositions which are to follow, or even a slight reading of the fragments themselves, will sufficiently show how thoroughly fantastic and arbitrary are his interpretations. Lassalle seems to have been misled partly by Hegel's misinterpretation of the passages from Aristotle noticed above, and partly by the principle of opposition which runs through a number of the sayings of Heraclitus—an opposition which, as we shall see later, was wholly physical, and far more simple than the abstruse logical meaning given it by Lassalle. This German scholar had no power or no wish to put himself in the attitude of the Greek mind, which was as widely different from his as possible. It was a mistake for this disciple of pure thought, bred in the stifling atmosphere of a nineteenth century Hegelian lecture-room, and powerless to transport himself out of it even in thought, to attempt to interpret the sentences of an ingenuous lover of Nature, who, five centuries before the Christian era, lived and moved in the free air of Ephesus. In this we do not mean to say that the philosophy of Heraclitus was purely physical rather than metaphysical, for we shall see that such was not the case, but primitive pre-Socratic metaphysics and the panlogism of Lassalle are as wide asunder as the poles. On this point, Benn, in the work already referred to, well says, "The Greek philosophers from Thales to Democritus did not even suspect the existence of those ethical and dialectical problems which long constituted the sole object of philosophical discussion" (Vol. 1, p. 4).

Those who wish to trace Lassalle's errors further may compare, on his mistaken conception of the Heraclitic fire, Zeller, Vol. 1, p. 591, 3[1]; Grote: Plato, Vol. 1, p. 33, note. On "the way up and down," compare Zeller, Vol. 1, p. 619, 1. On the flux, compare Schuster, p. 201; Zeller, Vol. 1, p. 577, 1.

The characterization of Lassalle's book as a whole

---

[1] The references to Zeller in the following pages are to the fourth German edition of Die Philosophie der Griechen.

is, that it is a striking example of great philosophic waste, turning as he does the rich and suggestive philosophy of the Ephesian into a wretched mouthful of Hegelian phrases. His citation of so many diverse sentences of Heraclitus, drawn from theology, ethics, nature, and man, and his discovery in all of them of his single ever-recurring notion of "die reine umschlagende Identität von Sein und Nichtsein," impresses us with the power which the tyranny of a single idea may have to so blur one's vision as to cause him to see that idea reflected in everything that is presented. It is not true, as Lassalle's motto goes, that there is no sentence of Heraclitus that Hegel has not incorporated in his Logic, but it is not far from the truth that there is no sentence of Heraclitus which Hegel and Lassalle have not either willfully or ignorantly perverted.

Schuster.

We will mention now the work of Paul Schuster (see above, p. 1). Schuster approaches the problem of the interpretation of Heraclitus with the advantage of a rich philological and historical knowledge. He suffers a disadvantage, however, in the magnitude of the task he undertakes, which is nothing less than the reconstruction of the order and plan of the book of Heraclitus itself. The interpretation of the fragments, he justly observes, depends upon the connection in which they occurred. It will be necessary, therefore, if we will grasp their true sense, to recover the plan of the original writing. Such a reconstruction Schuster holds to be possible, since by the law of selection, the fragments which have been preserved to us must have been the central thoughts of the original work. Contrary to Schleiermacher, he accepts as trustworthy the statement of Diogenes (Diog. Laert. ix. 5) that the book of Heraclitus was divided into three parts or Logoi, the first concerning "the all," the second political, the third theological. On this basis Schuster arranges the fragments, freely translated or rather paraphrased, and interspaced with the restored progress of thought. The well known obscurity of our philosopher, Schuster, contrary to all other critics except Teichmüller, supposes to have been partly, at least, intentional, as a precaution against persecution for atheism.[1]

The distinctive feature of Schuster's conception of Heraclitus is that he was not a distruster of the senses, but on the contrary the first philosopher who dared to base all knowledge upon sense experience. He was therefore the first of experimental philosophers. To this idea the introduction of Heraclitus' book was devoted. The majority of people, says the Ephesian, have little interest in that which immediately surrounds them, nor do they think to seek for knowledge by investigation of that with which they daily come in contact (Clement of Alex. Strom. ii. 2, p. 432; M. Aurelius iv. 46; cp. frags. 5, 93). Nevertheless, that which surrounds us is the source of knowledge. Nature is not irrational and dumb, but is an ever living Voice plainly revealing the law of the world. This Voice of Nature is the Heraclitic Logos. The thought which Heraclitus utters in the passage standing at the beginning of his book (frag. 2, Hippolytus, Ref. haer. ix. 9; cp. Aristotle, Rhet. iii. 5, p. 1407, b. 14) is no other than that which since the Renaissance has

---

[1] Compare Plutarch. Pyth orac. 21, p. 404; = frag. 11; Clement of Alex Strom v. 13, p 699, = frag 116. The numbers refer not to Schuster's numbering of the fragments, but to that of the present work, which is the numeration of Bywater.

inspired natural science and its accompanying speculation, namely, that truth is to be won by observation of the visible world. But the people, he complains, despise the revelation which Nature offers us with audible voice. Why, asks Heraclitus (Hippolytus, Ref. haer. ix. 9; cp. frag. 47), should an invisible harmony be better than a visible? It is not better, but, on the contrary, whatever is the object of seeing, hearing, or investigation, that I particularly honor (idem ix. 10; cp. frag. 13). Men, therefore, must trust their eyes (Polybius, xii. 27; cp. frag. 15) and not make reckless guesses concerning the weightiest things (Diog. Laert. ix. 73; cp. frag. 48). That Heraclitus' theory of knowledge, therefore, based it upon sense perception and reflection thereupon, is shown, continues Schuster, not only by the above passages, but also by the fact that the exaggerated form of the theory held by Protagoras (cp. Plato's Theaetetus) must necessarily have had its source in Heraclitus, his master. None the less is this shown also by Parmenides' attack on the empirical theory of knowledge (Sextus Empir. vii. 3), which could have been aimed only at the philosopher of Ephesus (Schuster, pp. 7 and 13–42).

Turning now from the theory of knowledge to its results, the first law which the observation of Nature teaches us is the law of eternal and recurrent motion ($\pi\acute{\alpha}\nu\tau\alpha\ \chi\omega\rho\epsilon\tilde{\iota}\ \varkappa\alpha\grave{\iota}\ o\grave{\upsilon}\delta\grave{\epsilon}\nu\ \mu\acute{\epsilon}\nu\epsilon\iota$, Plato, Crat. p. 402 A). The starting point and central position of our philosopher we must find in this recurrent motion, rather than in the primitive fire which itself held a subordinate place in the system. But the Heraclitic motion was not conceived as any absolute molecular change in the modern sense, nor yet as that absolute instability which appeared in the nihilism of the later

Heracliteans. It was rather conceived in a simpler way, as a general law that everything comes to an end and there is nothing permanent. Under this was included: 1) spacial motion, as of the flowing river; 2) qualitative change, as in the human body; 3) a kind of periodicity which brings everything under its dominion. The last was the most emphasized. Birth and death are universal; nothing escapes this fate. There is no fixed or unmoved being above or outside the shifting world, no divine heavenly existence that does not change, but all is involved in the same perpetual ebb and flow, rise and fall, life and death (Schuster, p. 81 ff.).

But this life and death of the universe is literal, not figurative. The world itself is a great living organism subject to the same alternation of elemental fire, air, and water. This thoroughgoing hylozoism which Schuster attributes to Heraclitus, he bases principally on the writing *de diaeta* of Pseudo-Hippocrates, who, he believes, made a free use of the work of Heraclitus, if he did not directly plagiarize from him. Comparing this writing (particularly the passage, c. 10, p. 638) with Plato's Timaeus (p. 40 A, also drawn from Heraclitus), he ventures to reconstruct the original as follows: "Everything passes away and nothing persists. So it is with the river, and so with mortal beings; in whom continually fire dies in the birth of air, and air in the birth of water. So also with the divine heavenly existence, which is subject to the same process, for we are in reality only an imitation of that and of the whole world; as it happens with that so it must happen with us, and inversely we may judge of that by ourselves" (Schuster, p. 118).

The life principle of the universe, as of the human

organism, is fire. This fire is everywhere present, so that "everything is full of gods and souls" (Diog. Laert. ix. 7). The life of the body is sustained by the breath which inhales the dry vapors kindred to fire. At night, when the sun is extinguished and the world becomes unconscious, we inhale the dark wet vapors and sink into death-like sleep (Schuster, p. 135).

The sun, which is new every day, changes at night into the surrounding air and then into the water of the sea. The sea produces the daily sun, as it is the source of all earthly phenomena. On a large scale this threefold change takes place with the universe, which will ultimately be consumed in fire, again to become sea and cosmos. This is "the way up and down"—not a circular movement of the elements within the cosmos (Zeller), but the periodicity of the world itself. The way up and the way down relate only to the cosmogony. The latter is the creation of the world by condensation of fire into water, then earth; the former is the reverse process of vaporization (Id. p. 169).

This law or order is not dependent upon any divine purposeful will, but all is ruled by an inherent necessary "fate." The elemental fire carries within itself the tendency toward change, and thus pursuing the way down, it enters the "strife" and war of opposites which condition the birth of the world ($\delta\iota\alpha\varkappa\acute{o}\sigma\mu\eta\sigma\iota\varsigma$), and experience that hunger ($\chi\rho\eta\sigma\mu\sigma\sigma\acute{u}\nu\eta$) which arises in a state where life is dependent upon nourishment, and where satiety ($\varkappa\acute{o}\rho\sigma\varsigma$) is only again found when, in pursuit of the way up, opposites are annulled, and "unity" and "peace" again emerge in the pure original fire ($\grave{\epsilon}\varkappa\pi\acute{u}\rho\omega\sigma\iota\varsigma$). This impulse of Nature towards change is conceived now as "destiny," "force," "necessity," "justice," or, when exhibited in definite forms of time and matter, as "intelligence" (Id. p. 182, 194 ff.).

The Heraclitic harmony of opposites, as of the bow and the lyre, is a purely physical harmony. It is simply the operation of the strife of opposite forces, by which motion within an organism, at the point where if further continued it would endanger the whole, is balanced and caused to return within the limits of a determined amplitude (Id. p. 230 ff.).

The identity of opposites means only that very different properties may unite in the same physical thing, either by simultaneous comparison with different things or successive comparison with a changeable thing (Id. pp. 236, 243).

The second or political section of Heraclitus' work treated of arts, ethics, society, and politics. It aimed to show how human arts are imitations of Nature, and how organized life, as in the universe and the individual, so in the state, is the secret of unity in variety. The central thought was the analogy existing between man and the universe, between the microcosm and the macrocosm, from which it results that the true ethical principle lies in imitation of Nature, and that law is founded on early customs which sprang from Nature (Id. p. 310 ff.).

The third or theological section was mainly devoted to showing that the names of things are designations of their essence. That Heraclitus himself, not merely his followers, held the φύσει ὀρθότης ὀνομάτων, and used etymologies as proofs of the nature of things, Schuster believes is both consistent with his philosophy and conclusively proved by Plato's Cratylus. Primitive men named things from the language which Nature spoke to them; names, therefore, give us the truth of things. Etymologies of the names of the gods was the proof first brought forward, as in Plato's Cratylus; hence the name of this section of the work. To show this connection of names and things was to prove the intimate connection of man with Nature, and so to lead to the conclusion that all knowledge is based on experience, which, indeed, was the end he had in view (Id. p. 317 ff.).

It is not our purpose to criticize in detail Schuster's conception of Heraclitus. Much of it will commend itself to the careful student of the remains, particularly that which relates to the Heraclitic flux and its relation to the primitive fire. Suggestive, also, if not unimpeachable, is his conception of the relation of the microcosm to the macrocosm, and of the harmony and identity of opposites. In his exposition of these doctrines, Schuster has rendered valuable service. We can by no means, however, allow thus tentatively to pass, Schuster's conception of Heraclitus as a purely empirical philosopher. Before noticing this, a word needs to be said in regard to Schuster's method as a whole. As to the latter, the very extent of the task proposed made over-systemization inevitable. In criticism of Schuster's attempt, Zeller has well said that with the extant material of Heraclitus' book, the recovery of its plan is impossible (Vol. 1, p. 570, note). Such a plan of reconstruction as that which Schuster undertakes, demands the power not only to penetrate the sense of every fragment, but also so to read the mind of the author as to be able to restore that of the large absent portions. The small number and enigmatical character of the fragments which are extant, together with the contradictory character of ancient testimony to Heraclitus, makes such a task extremely hazardous.

It can be carried through only by the help of "unlimited conjecture." Such conjecture Schuster has used extensively. The necessity of carrying through his plan has led him to find in some passages more meaning than they will justly bear, while his apparently preconceived notion as to the wholly empirical character of the system has led him to distort the meaning of many sentences. We shall see examples of this presently. Incidentally, his method may be illustrated by his connection and use of the two passages: ἀνθρώπους μένει ἀποθανόντας, ἅσσα οὐκ ἔλπονται οὐδὲ δοκέουσι (Clement of Alex. Strom. iv. 22, p. 630; cp. frag. 122), and αἱ ψυχαὶ ὀσμῶνται καθ' ᾅδην (Plutarch, de Fac. in orbe lun. 28, p. 943; cp. frag. 38). Schuster conjectures that these passages came together in the original work, and he renders and interprets them as follows : "There awaits men in death what they neither hope nor believe," namely, rest and the joy of a sleep-like condition (!), so that even instinctively "souls scent out death," desiring to obtain it (Schuster, p. 190). Not to speak of the forced translation of the latter fragment, only the most vivid imagination would think of using these passages in this way, especially as Clement himself, in his use of the first passage, refers it to the punishments which happen to men after death (see below, frags. 122 and 124, sources), and Plutarch, in respect to the second, uses it as proof that souls in Hades are nourished by vapors (see below, frag. 38, sources). But Schuster's conception of Heraclitus did not admit of belief in a distinct life after death, and it was necessary to make these passages fit in with the plan. The attempt to weave the fragments into a connected whole, and their division into the three Logoi, may be regarded on the whole as a decided failure.

Schuster finds only thirteen fragments for the concluding theological section, although our knowledge of Heraclitus and his times would rather indicate, as indeed Teichmüller thinks probable, that the theological section was the principal portion of the book.

Turning now to the theory of knowledge, according to Schuster, as we have seen, Heraclitus is an empiricist and sensationalist and knows no world but the visible. With this conclusion we cannot agree. Schuster's argument that this doctrine must have arisen with Heraclitus since it was held by Protagoras, his disciple, has little weight. The order of development was rather that pointed out by Plato himself in the Theaetetus (p. 151 ff.), namely, that the sensational theory of knowledge was the outcome of the Protagorean doctrine that man is the measure of all things, and that this in turn grew out of the Heraclitic flux. No doubt the sensational theory was implied by the Sophists, but it was incipient with them and not yet formulated. Much less can it be attributed to Heraclitus, whose contribution to the theory began and ended with the eternal flux. A sensational theory of knowledge, it is quite true, was likely to be an outcome of the Ephesian's philosophy, but he did not himself proceed thus far. The question, theoretically considered, was beyond his time. There are passages which indicate that he held, inconsistently it may be, quite the opposite doctrine. "Eyes and ears," he says, "are bad witnesses to men having rude souls" (Sextus Emp. adv. Math. vii. 126; =frag. 4; cp. frags. 3, 5, 6, 19, etc., and below (p. 50). The passage which offers Schuster the strongest support for his sensationalism is that noted above (p. 13) from Hippolytus, "Whatever concerns seeing, hearing and learning ($\mu\acute{a}\theta\eta\sigma\iota\varsigma$, Schuster translates "Erforschung"), I particularly honor" (frag. 13). Adopting the simplest and most natural meaning of this passage, it has no bearing on any theory of knowledge, but means merely, as Pfleiderer points out (Heraklit, p. 64, note), that Heraclitus prefers the pleasures of the higher senses, as of seeing, hearing, and the knowledge acquired thereby, to the sensual pleasures of the lower senses which the masses pursue. If, however, Schuster will take it in a theoretical sense, then it comes into conflict with the other passage, "The hidden harmony is better than the visible." The contradiction is foreseen by Schuster, who deliberately changes the latter into a question (see above, p. 13), without a shadow of right, as may be seen by reference to the context in Hippolytus (see below, frag. 47), who expressly states that the two passages seem to conflict. Further support for his interpretation Schuster seeks in the following passage from Hippolytus:

Τοῦ δὲ λόγου τοῦδ' ἐόντος αἰεὶ ἀξύνετοι γίνονται ἄνθρωποι καὶ πρόσθεν ἢ ἀκοῦσαι καὶ ἀκούσαντες τὸ πρῶτον. γινομένων γὰρ πάντων κατὰ τὸν λόγον τόνδε ἀπείροισι ἐοίκασι πειρώμενοι καὶ ἐπέων καὶ ἔργων τοιουτέων ὁκοίων ἐγὼ διηγεῦμαι, διαιρέων ἕκαστον κατὰ φύσιν καὶ φράζων ὅκως ἔχει (Ref. haer. ix. 9; = frag. 2).

This is the passage of which Schuster says that if Heraclitus had written nothing more it would have given him a place of honor in philosophy, for here for the first time appeared the thought that has inspired speculation and modern science since the Renaissance, that truth is to be sought in the observation of Nature. But we are unable to find here any such meaning. The sense of the passage depends upon the sense of Logos. Of course, if Schuster is free to translate this word in any way he chooses, he can get from the passage almost any meaning. He chooses to render it the Voice of Nature or the Speech of the visible world. In this he is not supported by any other critics. By ancient commentators of Heraclitus the Logos was understood as Reason, and in this general sense it is taken by modern commentators including Heinze, Zeller, Teichmüller, and Pfleiderer, although more specifically they see that, in harmony with the whole Heraclitic philosophy, it is to be taken as Reason immanent in the world as Order or Law. Schuster objects that Logos could not mean Reason, since before the time of Heraclitus it had never been so used, and no author would venture to introduce at the very beginning of his work words with new meanings. But precisely the same objection applies to its meaning the Speech of Nature, for the whole point in Schuster's exposition is that this was an original idea with Heraclitus. If the Logos is conceived as Order, this objection is met, since this meaning is given in the derivation of the word. Moreover, if Schuster could show that the word meant "speech" or "discourse," then the discourse referred to must have been not that of Nature but of the author himself. Finally, if we adopt Reason as the meaning of Logos here, the whole passage, so far from supporting, directly refutes Schuster's sensational theory of knowledge. Another argument for the empiricism of Heraclitus, Schuster seeks in his denunciation of the people for their failure to interest themselves in acquiring knowledge by empirical investigation of the things that surround them, which he bases on a couple of passages from Clement and M. Aurelius (see above, p. 12). Heraclitus, in fact, said nothing of the kind; but Schuster, by conjectural reconstruction of the text and an arbitrary translation, extracts a theoretical meaning from simple sentences which no one who had not a preconceived theory to support would ever imagine to mean more than a reproach upon the masses for their superficiality and neglect of interest in a deeper knowledge of the world (see Schuster, p. 17, and cp. frags. 5, 93). What Heraclitus' theory of knowledge really was we shall see more fully in the examination of Pfleiderer's position later. Here it is sufficient to add that, whatever empirical tendency his philosophy may have had, any such positive doctrine as that which Schuster ascribes to him was far beyond the time of Heraclitus.

Schuster's interpretation of the Heraclitic χρησμοσύνη and κόρος is also open to criticism. Zeller, indeed, has given a similar explanation of these words (Vol. 1, p. 641), but Pfleiderer has understood them differently (p. 176). From Heraclitus himself there remains only the two above words (frag. 24). Hippolytus (Ref. haer. ix. 10, cp. frag. 24, sources) says that the arrangement of the world (διακόσμησις), Heraclitus called "craving" (χρησμοσύνη), and the conflagration of the world (ἐκπύρωσις) he called "satiety" (κόρος). Schuster, therefore, understanding by διακόσμησις, not the process of world-building, that is, the passing of the homogeneous original fire into the manifold of divided existence, but the completed manifold world itself or the κόσμος, interprets the "craving" or hunger as belonging to the present differentiated world, which hungers, as it were, to get back into the state of original fire or satiety. The testimony is too meagre to say that this is not a possible interpretation, but it seems to be wrong. For Schuster admits, as of course he must, that the original fire carries within itself an impulse to change and develop into a manifold world. But this impulse to change is hardly consistent with a state of perfect "satiety." If now we take διαϰόσμησις in its primary signification denoting the action or process of arranging, then craving becomes the designation of the world-building process itself. Craving then is nothing but the original impulse to evolve itself, contained in the primitive fire, while the reverse process, the conflagration, is satiety, or better, the result of satiety.

TEICHMÜLLER.

The work of Teichmüller (see above, p. 1) does not profess to be a complete exposition of the philosophy of Heraclitus, but to indicate rather the direction in which the interpretation is to be found. Teichmüller believes that the philosophy of the ancients is to be interpreted by their theories of Nature. Physics came before metaphysics. Particularly does this apply to Heraclitus of Ephesus. His philosophy of Nature, therefore, is the key with which Teichmüller will unlock the secrets of his system (Teichmüller, I, p. 3).

But yet Heraclitus was not a naturalist. Of the sun, moon, eclipses, seasons, or earth, he has little to say. In the astronomy of Anaximander or the mathematics of Pythagoras he took little interest. On such polymathy he cast a slur (Diog. Laert. ix. 1; cp. frag. 16). He went back to Thales and started from his childlike conception of Nature. To Heraclitus the earth was flat, extending with its land and sea indefinitely in each direction. The sun, therefore, describes only a semicircle, kindled every morning from the sea and extinguished in it every evening. Moreover, the sun is no larger than it looks (Diog. Laert. ix. 7). The sun, therefore, cannot pass his boundaries (of the half-circle), else the Erinyes (who inhabit the lower world) will find him out (Plutarch, de Exil. ii. p. 604; = frag. 29). Up and down are not relative but absolute directions (Teichmüller, I, p. 14).

Thus upon physical grounds we may interpret at once some of the aphorisms. For instance, since the sun is a daily exhalation from the earth, sun and earth must have in part a common substance; hence Dionysus and Hades are the same (Clement of Alex. Protrept. ii. p. 30; cp. frag. 127), since the former stands for the sun and the latter for the lower world. Likewise day and night are the same (Hippolytus, Ref. haer. ix. 10; cp. frag. 35), since they are essentially of the same elements, the difference being only one of degree, the former having a preponderance of the light and dry, the latter of the dark and moist (Teichmüller, I, pp. 26, 56).

The four elements, fire, air, earth, and water, are not, as with Empedocles, unchangeable elements, but in ceaseless qualitative change are continually passing into one another. Experience itself teaches this in the daily observation of such phenomena as the drying up of swamps, the melting of solids, and the evaporation of liquids (Id. I, p. 58).

Fire is not a symbol, but is real fire that burns and crackles. It is the ground principle, the entelechy of the world, in which reside life, soul, reason. It is God himself. It is absolute purity. It rules in the pure upper air, the realm of the sun. Its antithesis is moisture, absolute impurity, which rules in the lower regions of the earth. The sun with his clear light moves in the upper fiery air. The moon with her dimmed light moves in the lower moister air. The central thought, therefore, is purification, or "the way up," from the moist and earthy to the dry and fiery (Id. I, p. 62 ff.).

The psychology of Heraclitus is not analogous, but identical with his physics. The soul is the pure, light, fiery, incorporeal principle which burns like the sun. Its degree of life and intelligence depends upon its purity from moisture. The stupid drunken man has a moist soul (Stobaeus Floril. v. 120 ; cp. frag. 73). "The dry soul is the wisest and best" (fråg. 74). In sleep the fire principle burns low ; in death it is extinguished, when the soul, like the sun at night, sinks into the dark regions of Hades. Hence it follows that there was with Heraclitus no doctrine of the immortality of the soul (Teichmüller, I, p. 74 ff.).

Ethics, therefore, is purification, and in this thought we see the origin of that general idea which as "Catharsis" became prominent in Plato and later philosophy. Teichmüller finds it of the greatest interest to have traced the history of this idea, with its related one of "separation" or "apartness," back to Heraclitus. "Of all whose words I have heard," says the latter, "no one has attained to this—to know that Wisdom is apart ($\gamma\varepsilon\chi\omega\rho\iota\sigma\mu\varepsilon\nu\text{o}\nu$) from all" (Stobaeus Floril. iii. 81 ; = frag. 18). This "separateness" of Wisdom, which was only another term for reason, God or pure fire, reveals the origin of the distinction of the immaterial from the material. With Heraclitus, to be sure, the idea of immateriality in its later sense was not present, but fire as the *most* incorporeal being of which he knew, identical with reason and intelligence, was set over against the crude material world. We have therefore here neither spiritualism nor crude materialism, but the beginning of the distinction between the two. With Anaxagoras another step was taken when fire was dropped and the Nous was conceived in pure separateness apart even from fire. Following Anaxagoras, Plato regarded the Ideas as distinct and separate (εἰλικρινές, κεχωρισμένον). In Aristotle it appears as the separation (χωριστόν) which belongs to absolute spirit or pure form. Finally in the New Testament it is seen as the purity (εἰλικρίνεια) which is opposed to the flesh (Paul, Epist. to Corinth. II, i. 12; ii. 17). Human intelligence, according to Heraclitus, attains only in the case of a few to this greatest purity, this highest virtue, this most perfect knowledge. They are the chosen ones, the elect (ἐκλεκτοί) (Teichmüller, I, p. 112 ff.).

The senses, since they partake of the earthy character of the body, give us only deceitful testimony as compared with the pure light of Reason, which alone, since it is of the essence of all things, that is, fire, has the power to know all. Here therefore was the first distinction of the intelligible from the sensible world (Id. I, p. 97).

Again, in the qualitative change of Heraclitus we discover the incipient idea of the actual and potential first formulated by Aristotle. Since the elements pass into one another, they must be in some sense the same. Water is fire and fire is water. But since water is not actually fire, it must be so potentially. To express this idea, Heraclitus used such phrases as "self-concealment," "sunset," "death," "sleep," "seed" (Id. I, p. 92 ff.).

Moreover, inasmuch as we have a progress from the potential to the actual, from the moist and earthy to the dry and fiery, that is, from the *worse* to the *better*, we find in Heraclitus the recognition of an end or purpose in Nature, or a sort of teleology, subject, however, to the rule of rigid necessity (Id. I, p. 137).

The flux of all things Teichmüller understands not

as a metaphysical proposition, but as a physical truth gained by generalization from direct observation of Nature. Furthermore, it was nothing new, all the philosophers from Thales on having taught the motion of things between beginning and end (Id. I, p. 121).

That which *was* new in this part of Heraclitus' work was his opposition to the transcendentalism of Xenophanes. Over against the absolute, unmoved and undivided unity of the Eleatic philosopher, Heraclitus placed the unity of opposition. In Xenophanes' system, above all stood the immovable, transcendent God. In Heraclitus' system there was nothing transcendent or immovable, but all was pursuing the endless way upward and downward. His God was ceaselessly taking new forms. Gods become men, and men gods (Heraclitus, Alleg. Hom. 24, p. 51, Mehler; cp. frag. 67). The immanent replaces the transcendent. Here emerges the historically significant idea of unity. Against the unity of Xenophanes, a unity opposed to the manifold, Heraclitus grasped the idea of a unity which includes the manifold within itself. "Unite whole and part, agreement and disagreement, accordant and disaccordant—from all comes one, and from one all" (Arist. de mundo 5, p. 396, b. 12 ; =frag. 59). Everywhere is war, but from the war of opposites results the most beautiful harmony (cp. frag. 46). Here three principles are involved: 1). Through strife all things arise; the birth of water is the death of fire, the death of water is the birth of earth, etc. (cp. frag. 68). 2). Through strife of opposites all things are preserved; take away one, the other falls; sickness is conditioned by health, hunger by satiety (cp. frag. 104). 3). There is an alternating mastery of one or the other opposite; hence it follows that since all opposites proceed from one another, they are the same (Teichmüller, I, p. 130 ff.).

What did Heraclitus mean by the visible and invisible harmony? Teichmüller censures Schuster for failing to recognize that most significant side of Heraclitus' philosophy which is represented by the invisible harmony—in other words, for reducing him to a mere sensationalist. The visible harmony, according to Teichmüller, is the entire sensible world, in which the war of opposites results in a harmony of the whole. But the invisible harmony is the divine, all-ruling and all-producing Wisdom or World-reason, concealed from the senses and the sense-loving masses and revealed only to pure intellect. Thus Heraclitus, to whom there was an intelligible world revealing itself to intellect alone, and in the recognition of which was the highest virtue, was the forerunner of Plato (Id. I, pp. 154, 161 ff.).

By the Logos of Heraclitus was indicated Law, Truth, Wisdom, Reason. It was more than blind law, thinks Teichmüller, it was self-conscious intelligence; for self-consciousness, according to Heraclitus, who praised the Delphic motto, "Know thyself," is the highest activity of man, and how could he attribute less to God, from whom man learns like a child? (cp. frag. 97). But this self-conscious reason is not to be understood as a constant, ever abiding condition. God, who in this purely pantheistic system is one with the world, is himself subject to the eternal law of ceaseless change, pursuing forever the downward and upward way. But is not then God, Logos, Reason, subject, after all, to some higher destiny (εἱμαρμένη)? No, says Teichmüller, for it is this very destiny which it is the highest wisdom in man to recognize, and which is, therefore, identical with the Wisdom which rules all. The difficulty here he so far admits, however, as to acknowledge that this doctrine is "dark and undetermined" (Id. I, p. 183 ff.).

Finally, says our author, there was no idea of personality of spirit in the philosophy of Heraclitus, as there was not in any Greek philosopher from Xenophanes to Plotinus (Id. I, 187).

In closing this part of his exposition, Teichmüller calls attention to the relation of Heraclitus to Anaxagoras. M. Heinze (Lehre vom Logos, p. 33), following Aristotle, attributes to Anaxagoras the introduction into philosophy of the idea of world-ruling intelligence. But, says Teichmüller, this idea was present to every Greek from Homer on. Its recognition by Heraclitus has been shown by the fact that everywhere he attributes to his God, wisdom ($\sigma o \varphi i a$), intelligent will ($\gamma \nu \omega \mu \eta$), reason ($\varphi \rho o \nu o \tilde{\upsilon} \nu$ and $\varphi \rho \varepsilon \nu \eta \rho \varepsilon \varsigma$), and recognized truth ($\lambda \acute{o} \gamma o \varsigma$). What then did Anaxagoras add? The history of the idea of transcendent reason turns upon two characteristics, Identity ($\tau a \upsilon \tau \acute{o} \tau \eta \varsigma$) and Pure Separation ($\varepsilon i \lambda i \varkappa \rho i \nu \acute{\varepsilon} \varsigma$). With Heraclitus both failed; the former, because the World Intelligence took part in the universal change; the latter, because it was mingled with matter. For, in choosing fire for his intelligent principle, although as Aristotle says he chose that which was least corporeal ($\dot{a} \sigma \omega \mu a \tau \dot{\omega} \tau a \tau o \nu$), he did not escape a sort of materialism. The new that Anaxagoras added, therefore, was the complete separation of reason from materiality. In a word, while the Logos of the Ephesian was at once world-soul and matter in endless motion, the Nous of Anaxagoras was motionless, passionless, soulless and immaterial. Identity, the other attribute, was added ·in the epoch-making work of Socrates when the content of reason was determined by the definition, following whom Plato established the complete transcendence of the ideal world (Teichmüller, I, 189 ff.).

Heraclitus assumed a world-year or world-period, the beginning of which was the flood, and whose end was to be a universal conflagration, the whole to be periodically repeated forever. In this he was preceded by Anaximander and followed by the Stoics. This general idea was adopted by the Christian Church, but the latter limited the number of worlds to three, the first ending with the flood; ours, the second, to end with the conflagration of the world; the third to be eternal (Epist. Pet. II, iii. 4 ff.; Clement of Rome, Epist. to Corinth. i. 57, 58); (Teichmüller, I, 198 ff.).

In the second part of his work, Teichmüller enters upon an exhaustive argument to show the dependence of the Heraclitic philosophy upon Egyptian theology. Heraclitus moved within the sphere of religious thought. He praised the Sibyl and defended revelation and inspiration (Plutarch, de Pyth. orac. 6, p. 397; cp. frag. 12). His obscure and oracular style, like that of the king at Delphi (cp. frag. 11), was in conformity with his religious character. Observation of Nature he fully neglected, depending for his sources more than any other philosopher upon the beliefs of the older theology. Without deciding how far Heraclitus is directly, as a student of the Book of Death, or indirectly by connection with the Greek Mysteries, dependent upon the religion of Egypt, he proceeds to indicate the interesting points of similarity between them (Teichmüller, II, p. 122).

Among the Egyptians the earth was flat and infinitely extended. The visible world arose out of water.

The upper world belonged to fire and the sun. As the sun of Heraclitus was daily generated from water, so Horus, as Ra of the sun, daily proceeded from Lotus the water. As the elements with Heraclitus proceed upward and downward, so the gods of the elements upon the steps in Hermopolis climb up and down (Id. II, p. 143).

With these illustrations, it is sufficient to say, without following him further in detail, that Teichmüller carries the comparison through the whole system of Heraclitus, and parallels his actual and potential, his unity of opposites, his eternal flux, strife, harmony, purification, Logos, and periodicity of the world, with similar notions found in the religion of Egypt.

In order to appreciate the worth of Teichmüller's work, it is necessary to remember that, as we have said, it does not profess to be a unified exposition of Heraclitus' philosophy, but a contribution to the history of philosophic ideas in their relation to him. In affording this service to the history of ideas, he has thrown a good deal of light upon the true interpretation of the philosophy of Heraclitus. But the very purpose of his task has caused him to put certain of the ideas into such prominence, that unless we are on our guard, we shall not get therefrom a well proportioned conception of the system as a whole. We shall do well, consequently, to make a short examination of the work outlined in the foregoing pages, to put the results, if we can, into their fit relation to the whole.

Concerning Teichmüller's starting point, namely, that the physics of Heraclitus is the key to his whole thought, we must observe, in passing, the inconsistency between the first part of Teichmüller's book, where this principle is made the basis of interpretation, and the second part, where it sinks into comparative insignificance when he discovers that Heraclitus is primarily a theologian and gets his ideas from Egyptian religion. To say that we shall better appreciate a philosopher's position if we understand his astronomy and his theories of the earth and nature, is of course true to every one. Moreover, that Heraclitus considered the earth as flat, the sun as moving in a semicircle and as no larger than it looks, the upper air as drier than the lower, and the lower world as dark and wet, there is no reason to deny. In fact, this cosmology, as Teichmüller details it, is so simple and blends so well with the Heraclitic sayings in general, that the picture of it once formed can hardly be banished from the mind. But that it adds much to the explication of the philosophy as a whole is doubtful. It is true that physics came before metaphysics, if by that is meant that men speculated about Nature before they speculated about being. But this distinction has little bearing on the interpretation of Heraclitus. A principle more to the point, and one that Teichmüller has not always observed, is that religion, poetry and metaphor came before either physics or metaphysics. From the very fact, also, that physics came before metaphysics, when the latter did come, men were compelled to express its truths in such physical terms as they were in possession of. He therefore who will see in the sentences of Heraclitus nothing beyond their physical and literal meaning, will miss the best part of his philosophy. For instance, Teichmüller interprets the saying that day and night are the same, as meaning that they are made up of the same physical constituents (see above, p. 24). If possible, this is worse than

Schuster's explanation that they are the same because they are each similar divisions of time (!), an explanation which Teichmüller very well ridicules (Id. I, p. 49). No such childish interpretations of this passage are necessary when it is seen that this is simply another antithesis to express Heraclitus' great thought of the unity of opposites, on the ground that by the universal law of change, opposites are forever passing into each other, as indeed is said in so many words in a passage from Plutarch which these critics seem to have slighted (Consol. ad Apoll. 10, p. 106; see frag. 78). Equally unnecessary and arbitrary is Teichmüller's singular attempt to prove on physical grounds the identity of the two gods, Dionysus and Hades (see above, p. 24).

In pursuance of his method, Teichmüller supposes that the Heraclitic fire was real fire such as our senses perceive, fire that burns and crackles and feels warm. No other critic agrees with him in this. Zeller especially opposes this conception (Vol. I, p. 588). It is not to be supposed that Teichmüller understands Heraclitus to mean that the present world and all its phenomena are real fire. Fire he conceives to be, rather, the first principle or $ἀρχή$, the real essence of the universe, chosen as water was by Thales or air by Anaximenes, only with more deliberation, since fire has the peculiarity of taking to itself nourishment. In a word, since anybody can see that our present earth, water, and air, are not fire that burns and crackles, all that Teichmüller can mean is that this kind of fire was the original thing out of which the present world was made. But there is not the least support for this meaning in any saying of Heraclitus. In all the sentences, fire is conceived as something of the present, something directly involved in the ceaseless change of the world. "Fire, (*i. e.*, κεραυνός, the thunderbolt)," he says, "rules all" (Hippolytus, Ref. haer. ix. 10; =frag. 28). "This world, the same for all, neither any of the gods nor any man has made, but it always was, *and is*, and shall be, an ever living fire" (Clement of Alex. Strom. v. 14, p. 711; =frag. 20). "Fire is exchanged for all things and all things for fire" (Plutarch, de EI. 8, p. 388; =frag. 22). These passages are sufficient to show that Teichmüller's conception of the fire is untenable. We may, however, mention the fact noted by Zeller (Vol. I, p. 588), that both Aristotle (de An. i. 2, 405, a, 25) and Simplicius (Phys. 8, a) explain that Heraclitus chose to call the world fire "in order to express the absolute life of Nature, and to make the restless change of phenomena comprehensible."

Another point that demands criticism is the idea of actuality and potentiality which Teichmüller finds hidden in Heraclitus' philosophy and metaphorically expressed by sunset, death, sleep, etc. Since there is a qualitative interchange of the elements, they must be in some sense the same. Water is fire and fire is water. But since water is not actually fire, it must be so potentially. Therefore, water is potential fire. Such is Teichmüller's reasoning, as we have seen. Of course, it can be reversed with equal right. Since fire is not actually water, it must be so potentially. Therefore, fire is potential water. Which is to say that we have here a simple reversible series in which there is not only an eternal progress (or regress) from fire to water, but equally, and under the same conditions, an eternal regress (or progress) from water to fire. Either, therefore, may, with as good right as the other, represent actuality or potentiality. In other words, actuality and potentiality are superfluous ideas in this system. In fact, this antithesis has no place in metaphysics outside the philosophy of Aristotle, and he who has failed to see that right in this connection lies the main difference between the philosophy of Aristotle and that of Heraclitus, has missed the most vital part of the latter. With Aristotle there is an eternal progress but no regress. The potential is ever passing into the actual, but not the reverse. To be sure, a thing may be both actual and potential, but not as regards the same thing. The hewn marble is potential as regards the statue and actual as regards the rough marble, but of course the hewn marble and the statue cannot be reciprocally potential or actual. Matter is eternally becoming form, but not the reverse. Thus follows Aristotle's necessary assumption of a prime mover, an inexhaustible source of motion, itself unmoved—pure actuality, without potentiality. Hence the mainspring of the peripatetic philosophy is the *unmoved moving first cause.* But the philosophy of the Ephesian is the reverse of all this. With him there is no fixed being whatever (see Teichmuller himself, I, p. 121 : " Es bleibt dabei nichts Festes zurück," etc.), no unmoved first cause outside the shifting world which is its own God and prime mover. Thus Teichmüller, in identifying the Heraclitic fire with the Aristotelian pure actuality, overlooked the slight difference that while the one is absolute motion, the other is absolute rest ! We are glad, however, not to find this Aristotelian notion, which, though prevalent in metaphysics, has never added a ray of light to the subject, present in the philosophy of the Ephesian, and we see here another case of over-interpretation by which

Heraclitus' innocent use of such terms as sunset, death, and self-concealment, caused Aristotelian metaphysics to be forced upon him.

In tracing the history of ideas, much emphasis has been laid by Teichmüller, as we have seen, upon the idea of purification (κάθαρσις) as it appears in Heraclitus, and in connection therewith he has found the beginning of the idea of the "apartness" or "separation" of the immaterial world, an idea so enormously enlarged by Anaxagoras and Plato. As regards the Catharsis proper, Teichmüller has rendered a service by pointing out Heraclitus' connection with the idea; but in reading Teichmüller's book, one would be easily led to believe that the Catharsis idea is much more prominent in Heraclitus than it really is, and as regards the doctrine of "separation," it seems at once so incongruous with the system as a whole that we must inquire what foundation, if any, there is for it. The student of Heraclitus knows, although the reader of Teichmuller might not suspect, that the words κάθαρσις, καθαρός, εἰλικρινές, εἰλικρίνεια, χωριστόν, χωρισθέν, ἐκλεκτοί, themselves do not occur in the authentic remains of his writings. One exception is to be noted. The word κεχωρισμένον occurs in the passage from Stobaeus already noticed (see above, p. 25). It is as follows: ʽΟκόσων λόγους ἤκουσα οὐδεὶς ἀφικνέεται ἐς τοῦτο, ὥστε γινώσκειν ὅτι σοφόν ἐστι πάντων κεχωρισμένον (Stobaeus Floril. iii. 81). This passage Teichmüller uses as his text in establishing the connection of Heraclitus with the doctrine of "separation," unfortunately, however, first because he has not found the correct interpretation of it, and second, because, if he had, it would stand in direct contradiction to the doctrine of immanence which he spends all the next chapter in estabishing for Heraclitus. Σοφόν in this passage does not stand for the world-ruling Wisdom or Reason, or Divine Law, of which Heraclitus has so much to say in other passages. To assert the "apartness" of that Law would be to disintegrate the entire system, the chief point of which is the immanence of the Divine Law as the element of *order* in the shifting world. It does not follow that because τὸ σοφόν is used in the above larger sense in the passage from Clement of Alexandria (Strom. v. 14, p. 718; = frag. 65), that σοφόν cannot be used in quite the ordinary sense in the present passage. That it is so is attested by the agreement of Schuster (p. 42), Heinze (Lehre vom Logos, p. 32), Zeller (Vol. I, p. 572, 1), and Pfleiderer (p. 60). Lassalle, indeed, agrees with Teichmüller. Schuster, following Heinze, understands the sentence to mean merely that wisdom is separated from all (men), that is, true wisdom is possessed by no one. Zeller, followed by Pfleiderer, renders it: "No one attains to this—to understand that wisdom is separated from all things, that is, has to go its own way independent of general opinion." Schuster's interpretation is the most natural, so that the fragment belongs among the many denunciations of the ignorance of the common people—as indeed Bywater places it—and has nothing to do with any theory of the "separateness" of an absolute or immaterial principle. Neither is there any other passage which supports this doctrine. In further support, however, of the Catharsis theory in general, Teichmüller alleges the passage from Plutarch (Vit. Rom. 28), which speaks of the future purification of the soul from all bodily and earthy elements, and which Teichmüller thinks to have a strong Heraclitic coloring. In this passage Heraclitus is quoted as saying that "the dry soul is the best," but beyond this fragment it is a mere conjecture that it was taken from him. The passage at any rate is unimportant. What then remains to establish any connection whatever of Heraclitus with the "history of the idea of the εἰλικρινές"? Only the most general antithesis of fire and moisture, with the added notion that the former is the better and the latter worse. Since the divine essence of the universe itself is fire, the way upward from earth and water to fire is the diviner process, and pure fire is the noblest and highest existence. But this is shown better in the ethical sphere. The soul itself is the fiery principle (Arist. de An. i. 2, p. 405, a, 25). "The dry soul is the wisest and best" (frag. 74). The soul of the drunken, stupid man is moist (cp. frag. 73). The highest good was to Heraclitus the clearest perception, and the clearest and most perfect perception was the perception of the Universal Law of Nature, the expression of which was pure fire; and such perception was coincident with that condition of the soul when it was most like the essence of the universe. This is the sum-total of the idea of the Catharsis found in Heraclitus. It is worthy of notice, to be sure, but it is not so different from what might be found in any philosophy, especially an ethical philosophy, as to make it of any great moment, either in the history of ideas or in the exposition of this system.

We have studied now those parts of Teichmüller's work which, either by reason of their incompleteness or manifest error, most needed examination, namely his method, his wrong conception of the Heraclitic fire, his useless and unfounded theory of the actual and potential and of the separateness of the immaterial, and his over-emphasized doctrine of the Catharsis. Concerning the other points, it is only necessary in addition to call attention to the extreme value of his contribution in his explanations of the relation of Heraclitus to Xenophanes, to Anaxagoras and to Plato, of the Heraclitic Logos, of the flux, of the unity of opposites, and of the invisible harmony and the intelligible world defended against the sensationalism of Schuster. In the second part of his work also, though its value is less, he has contributed not a little light by his emphasis of the theological character of this philosophy, though one doubts whether his laborious collection of resemblances between the philosophy of the Ephesian and the religion of Egypt has shed much light on Heraclitus' position. It is seen at once that by taking such general conceptions as war and harmony, purification, periodicity of the world, etc., it would be easy to make a long list of parallelisms between any religion and any system of philosophy not separated farther in time and place than Heraclitus of Ephesus and the Egyptians. The resemblances, however, are certainly not all accidental, but they are such as do not affect the originality of the Ephesian, and unfortunately do not add much to a better knowledge of his philosophy.

PFLEIDERER.

Dr. Edmund Pfleiderer comes forward in a recent volume of 380 pages (see above, p. 1), with an attempt to interpret the philosophy of Heraclitus from a new and independent standpoint. He expresses dissatisfaction with all previous results. Other critics have made the mistake of starting not from the positive but from the negative side, namely, from the universal flux (as Zeller), or from the law of opposites (as Lassalle). But the hatred of the opinions of the masses which

Heraclitus exhibits, calls for some greater philosophical departure than the above negative principles, which indeed were already well known truths. Moreover, if we take these for his starting point, we can get no consistent system, for the doctrine of the universal flux does not lead naturally to the law of opposites, but rather the reverse. Again, neither the flux nor the law of opposites harmonizes with the doctrine of fire. Finally, the pessimistic, nihilistic tendency of the theory of absolute change does not agree well with the deep rationality and world-order which Heraclitus recognizes in all things, nor with his psychology, eschatology, and ethics (Pfleiderer, p. 7 ff.).

We must look elsewhere for his ground principle. To find it, we must discover the genesis of this philosophy, which did not spring into being spontaneously, like Pallas Athena from the head of her father. It could not have come from the Eleatics, for the chronology forbids, nor from Pythagoras, whom Heraclitus reviles, nor finally from the physicists of Miletus, with whose astronomy Teichmüller has well shown our philosopher to be unacquainted. Its source is rather to be sought in the field of religion, and particularly in the Greek Mysteries. In the light of the Orphico-Dionysiac Mysteries, in a word, according to Pfleiderer, this philosophy is to be interpreted. Here is the long-sought key. The *mystic* holds it, as indeed Diogenes Laertius says:

Μὴ ταχὺς Ἡρακλείτου ἐπ' ὀμφαλὸν εἶλεε βίβλον
τοὐφεσίου· μάλα τοι δύσβατος ἀτραπιτός.
ὀρφνὴ καὶ σκότος ἐστὶν ἀλάμπετον· ἢν δέ σε μύστης
εἰσαγάγῃ, φανεροῦ λαμπρότερ' ἠελίου.—ix. 16.

With the religion of the Mysteries, in its older and purer form, Heraclitus was in full sympathy. By his family he was brought into close connection with it. Ephesus, too, his city, was a religious centre. Diogenes (ix. 6) relates that he deposited his book in the temple of Artemis. Heraclitus, indeed, was not a friend of the popular religion, but that was because of its abuses, and it was in particular the popular Olympian religion that he attacked. The connection of the Ephesian with the Mysteries may be considered as a deep-seated influence which their underlying principles exerted upon him. These religious principles he turned into metaphysics. His system as a whole was religious and metaphysical (Pfleiderer, p. 32 f.).

With this introduction, Pfleiderer proceeds as follows. Heraclitus' starting point lay positively in his *theory of knowledge,* which was a doctrine of speculative intuition and self-absorption. In this sense our author understands the fragment from Plutarch (adv. Colot. 20, p. 1118; = frag. 80), Ἐδιζησάμην ἐμεωυτόν, "I searched within myself," that is, I wrapped myself in thought, and so in this self-absorption I sought the kernel of all truth. Hence his contempt for the masses who act and speak without insight. But does not this conflict with those Heraclitic sentences which place the standard of truth and action in the common or universal (ξυνόν)? (cp. frags. 92, 91). Do these not lead as Schuster holds, to the rule, *Verum est, quod semper, quod ubique, quod ab omnibus creditum est?* No, says Pfleiderer, the common here does not mean the general opinion of the majority. All such interpretations are sufficiently refuted by that other passage, "To me, one is ten thousand if he be the best" (frag. 113). What Heraclitus really meant by the common (ξυνόν) was "the true inward universality." Absorption into one's inner self was absorption into that ground of reason which is identical with the divine principle of the world. By this universal reason under which he contemplated all things, Heraclitus meant nothing different from what by Spinoza was expressed by "sub specie aeternitatis," and in subsequent philosophy by "intellectual intuition" and "the standpoint of universal knowledge." Heraclitus fell back upon that universal instinct which in the form of human language is exhibited as the deposit of successive ages, and which again he did not distinguish from the voice of the Sibyl, representative of divine revelation. As respects the source of knowledge, Heraclitus as little as Spinoza, Fichte and Hegel, looked to himself as individual, but rather to that singular and qualitative divine source in which the individual participates (Pfleiderer, p. 46 ff.).

The senses, though they do not give us the whole truth, yet furnish the sufficient data that are to be interpreted by the light of reason. The errors of the masses do not arise from trusting the senses, for the latter give not a false, but a partial account. Their error lies in missing the spiritual band which unites the manifold of sense into the higher unity, an error distinctive of the popular polytheism as against the religion of the Mysteries (Id. p. 70).

The theory of knowledge, Heraclitus' starting point, being thus disposed of, Pfleiderer proceeds to discuss the material principles of his philosophy in their abstract metaphysical form. The keynote here is the *indestructibility of life.* The oscillating identity of life and death, a truth adopted from the Mysteries, is taken up by Heraclitus and elevated into a universal and metaphysical principle. It is based on the simple observation of Nature, which sees the life and light and warmth of summer passing into the death and darkness and cold of winter, only to be revived and restored in the never-failing spring. So on a smaller scale, day passes into night, but night ever again into day. So everywhere in Nature nothing passes away but to revive again. From this follows the hope of the universality of this law, the indestructibility of human life, and the resolution of the opposition between the light, warm life here above and the dark, cold death below. This is the hopeful element which characterizes the philosophy of the Ephesian. Over against it was the hopeless creed of the masses, whose complaint over the inexorable destiny of death found expression from the earliest times in the despairing lines of the poets. The common view does not see too much continuance and constancy in reality, but too little. "What we see waking," says Heraclitus, "is death, what we see sleeping is a dream" (Clement of Alex. iii. 3, p. 520; = frag. 64). Which means, that like the unreality and inconstancy of dreams is this ephemeral and perishing existence which we, the vulgar people, see when awake. Reversing this gloomy view, the Mysteries taught that Hades and Dionysus were the same (cp. frag. 127). That is, the god of death feared in the world below, is identical with the god of life and joy of the world here above, which is to say that the regenerative power of life persists even in death and shall overcome it (Pfleiderer, p. 74 ff.).

From this theory of the indestructibility of the fire force of life, Heraclitus passes to the ancillary truth of the unity of opposition in general. Hence he asserted the identity of day and night, winter and summer, young and old, sleeping and waking, hunger and satiety (cp. frags. 36, 78). His whole theory of the harmony of opposites was, as it were, apologetic. If life rules in death, why does death exist? It was in answer to this question that Heraclitus developed his science of opposition and strife, by showing the presence here of a general law (Pfleiderer, p. 84 ff.).

In the same spirit Pfleiderer interprets the much contested figure of the harmony of the world as the harmony of the bow and the lyre (see frags. 45, 56). Without rejecting the interpretation suggested by Bernays (Rhein. Mus. vii. p. 94) and followed by most other critics, which refers the figure to the *form* of the bow and of the lyre, their opposite stretching arms producing harmony by tension, Pfleiderer finds in the comparison still another meaning. The bow and the lyre are both attributes of Apollo, the slayer and the giver of life and joy. Thus the harmony *between* the bow and the lyre, as attributes of one god—symbols respectively of death and of life and joy—expresses the great thought of the harmony and reciprocal interchange of death and life (Pfleiderer, p. 89 ff.).

The Heraclitic flux of all things, says Pfleiderer, was not antecedent to his abstract teachings, but the logical consequence thereof. The identity of life and death led him to the identity of all opposites. But opposites are endlessly flowing or passing into each other. Hence from the principle that everything is opposition, follows the principle that everything flows. The universal flux is only a *picture* to make his religious metaphysical sentences intelligible (Id. p. 100 ff.).

The Heraclitic fire is real fire as opposed to the logical symbol of Lassalle, but not the strictly sensible fire that burns and crackles, as Teichmüller supposes. It is rather a less definite conception, which is taken now as fire, now as warmth, warm air or vapor. It is the concrete form or intuitional correlate of the metaphysical notion of life (Id. p. 120 ff.).

"The way up and down" refers not only to the transmutations of fire, water, and earth, but holds good in general for the oscillation of opposites, and particularly for the polarity of life and death (Id. p. 140).

As one result of his investigation, Pfleiderer affirms a strong optimistic element in the philosophy of the Ephesian. He contests the opinion of Schuster and Zeller that the endless destruction of single existences is kindred to the pessimistic doctrine of Anaximander, of the extinction of all individuals as an atonement for the "injustice" of individual existence. The process indeed goes on, but it has a bright side, and it is this that Heraclitus sees. Life, to be sure, is ever passing into death, but out of death life ever emerges. It is this thought, the powerlessness of death over the indestructible fire force of life, which Heraclitus emphasizes (Id. p. 180 ff.).

Still more decided is his rational optimism, his unswerving belief in a world well ordered and disposed. A deep rationality characterizes the universe (cp. frags. 2, 1, 91, 92, 98, 99, 96, 19). To express this idea, Heraclitus used the word Logos, which after his time played so prominent a part in the older philosophy. This word, passing even beyond its signification of "well ordered relation," conveyed finally with Heraclitus, as λόγος ξυνός, rather the idea of Reason immanent in the world (Pfleiderer, p. 231 ff.).

In the invisible harmony we find the same general thought. As distinguished from the visible harmony, which meant that external order of Nature insuring to the trustful peasant the never failing return of summer and winter, heat and frost, day and night,—the invisible harmony was that all-embracing harmony which is revealed to thought as the rational union of all oppositions. Against this theodicy there is no valid objection to be derived from the accounts which represent the Ephesian philosopher as sad and complaining, nor from the passages descriptive of the evils of life and the weakness of men (cp. frags. 86, 55, 112, etc.). In all cases these refer not to the philosopher's own opinions, but to the errors of the ignorant masses (Pfleiderer, p. 235 ff.).

The future existence of the soul, though not consistent with his physics and metaphysics, was nevertheless held from the religious and ethical standpoint. In fact it was involved, as has been shown, in Heraclitus' point of departure, so that we have less reason to complain of inconsistency in his case than we have, in reference to the same matter, in the case of the Stoics later (Id. p. 210).

We have given, perhaps, more space to the exposition of Pfleiderer's work than it relatively deserves, because it is the last word that has been spoken on Heraclitus, because, also, it has deservedly brought into prominence the optimism and the religious character of his philosophy, and because finally it presents another instructive example of over-systemization. It claims our attention, too, because the view it proposes is a complete reversal of the prevalent conception of Heraclitus, and if seriously taken, changes the whole tenor of his philosophy.

In what follows we shall examine chiefly the two main points in Pfleiderer's work, namely, the theory of knowledge and the connection with the Greek Mysteries; the latter, because it is Pfleiderer's particular contribution, and the former, because it will open to us an important aspect of the Ephesian's philosophy.

In the first place, however, it can by no means be admitted that the doctrine of the flux and the harmony of opposites represent the negative side of his system, and are secondary to his theory of knowledge and his religious dogmas. The unanimous testimony of the ancients cannot be thus easily set aside. That of Plato and Aristotle alone is decisive. Pfleiderer objects that Plato's purpose, which was to establish the changelessness of noumena against the change of phenomena, led him to emphasize the flux of Heraclitus. But if Heraclitus' positive teachings were, as Pfleiderer says, first of all the theory of knowledge, this and not the flux must have been emphasized in the Theaetetus where the theory of knowledge was Plato's theme. It is sufficient, however, here to note that what Heraclitus has stood for in philosophy from his own time to the present, is the doctrine of absolute change, and this doctrine may, therefore, properly be called the positive side of his philosophy. If what Pfleiderer means is that the theory of knowledge and not the flux was his starting point, he would have a shadow more of right. It is, however, misleading to say that his theory of knowledge was his starting point, for, as we have indicated in our examination of Schuster's work, Heraclitus was not concerned with a theory of knowledge as such. To state in a word what his point of departure really was, regarded from a common-sense view, it was his conviction that he was in possession of new truth which the blindness and ignorance of men prevented them from seeing (the point of departure indeed of almost every one who writes a book), and the three leading ideas in this new truth were: 1. the absence of that stability in Nature which the untrained senses perceive; 2. the unsuspected presence of a universal law of order; 3. the law of strife which brings unity out of diversity. In one sense this may be called a theory of knowledge, and only in this sense was it his starting point.

But concerning the theory of knowledge itself, we cannot accept Pfleiderer's position. By placing it in speculative intuition and self-absorption, he has rushed to the very opposite extreme of Schuster's sensationalism, and in so doing has equally misrepresented Heraclitus. Either extreme is forcing a modern theory of knowledge upon the Ephesian of which he was wholly innocent. What support has Pfleiderer for his "self-absorption" theory? None whatever. He alleges the fragment 'Εδιζησάμην ἐμεωυτόν (cp. frag. 80), which he arbitrarily renders, "I searched within myself" ("Ich forschte in mir selbst"). This fragment is from Plutarch (adv. Colot. 20, p. 1118), Diogenes Laertius (ix. 5; cp. frag. 80, sources), and others. Plutarch understands it to refer simply to self-knowledge like the Γνῶθι σαυτόν at Delphi (similarly Julian, Or. vi. p. 185 A). Diogenes understands it as referring to self-instruction (similarly Tatian, Or. ad Graec. 3). Diogenes says, "He (Heraclitus) was a pupil of no one, but he said that he inquired for himself and learned all things by himself" (ἤκουσέ τ' οὐδενός, ἀλλ' αὐτὸν ἔφη διζήσασθαι καὶ μαθεῖν πάντα παρ' ἑωυτοῦ). The latter seems to be its true meaning, as is seen by comparing the passage from Polybius (xii. 27; cp. frag. 15), "The eyes are better witnesses than the ears." As here he means to say that men should see for themselves and not trust to the reports of others, so in the fragment in question he means only that he himself has inquired of himself and not of others (cp. also frags. 14, 13). But Pfleiderer, in order to support a theory, has taken these two innocent words and pressed them into a doctrine of contemplative intuition, by giving them the meaning, "I wrapped myself in thought" ("Ich versenkte mich sinnend und forschend," etc., p. 47). So far is it from the case that Heraclitus sought the source of knowledge by turning inward, that he expressed himself directly to the contrary. Thus we read in Plutarch (de Superst. 3, p. 166; = frag. 95): ὁ Ἡράκλειτός φησι, τοῖς ἐγρηγορόσιν ἕνα καὶ κοινὸν κόσμον εἶναι, τῶν δὲ κοιμωμένων ἕκαστον εἰς ἴδιον ἀποστρέφεσθαι, the sense of which is well given by Campbell (Theaetetus of Plato, p. 246), "To live in the light of the universal Order is to be awake, to turn aside into our own microcosm is to go to sleep." Again, the whole passage from Sextus Empiricus (adv. Math. vii. 132, 133; cp. frags. 92, 2) is conclusive. "For," says Sextus, "having thus statedly shown that we do and think everything by participation in the divine reason, he [Heraclitus] adds, 'It is necessary therefore to follow the common, for although the Law of Reason is common, the majority of people live as though they had an understanding of their own.' But this is nothing else than an explanation of the mode of the universal disposition of things. As far therefore as we participate in the memory of this, we are true, *but as far as we separate ourselves individually we are false.* A more express denial of any self-absorption or *a priori* theory of knowledge would be impossible. Heraclitus is constantly urging men to come out of themselves and place themselves in an attitude of *receptivity* to that which surrounds them, and not go about as if self-included (cp. frags. 94, 3, 2). But what does Heraclitus mean by participation in the divine or universal Reason? Is not this just Pfleiderer's position when he says that the Ephesian as little as Fichte or Hegel looked to himself as individual, but rather to that absolute reason in which the individual participates? The difference is radical and vital, but Pfleiderer, like Lassalle, failed to see it because he did not free himself from strictly modern theories of knowledge. The difference is simply this. The universal reason of which Pfleiderer is speaking is that in which man *necessarily and by his intellectual nature* participates. That of Heraclitus is the divine Reason, in which man *ought* to participate but may not. Pfleiderer's universal reason is universal *in* man. That of Heraclitus, outside of and independent of man. The latter, so far from being necessarily involved in thought, is independent of thought. It is that pure, fiery and godlike essence, the apprehension of which gives rationality in the measure in which it is possessed. No reader, therefore, who can think of only two theories of knowledge, a strictly *a priori* theory and a strictly empirical theory, can understand Heraclitus. But, it may be asked, if knowledge does not come from without through the senses, nor from within from the nature of thought, whence does it come? Heraclitus, however, would not be disturbed by such a modern dilemma. There is reason, in fact, to believe, though it sounds strange to us, that he supposed this divine rational essence to be inhaled in the air we breathe (cp. Sextus Emp. adv. Math. vii. 127, 132). It exists in that which surrounds us ($περιέχον$), and the measure of our rationality depends on the degree in which we can possess ourselves of this divine flame. There was no conciseness of thought here, however, and Heraclitus seemed to think that it was partly apprehended through the senses, that is, the most perfect condition of receptivity to truth was the condition in which a man was most *awake*. The stupidest man is he who is asleep, blind, self-involved, and we may add, self-absorbed (cp. frags. 95, 90, 77, 3, 2, 94). Hence, if we have rightly interpreted Heraclitus here, a man might wrap himself in thought forever and be no nearer to truth. The source of knowledge did not lie in that direction to any pre-Socratic Greek philosopher. Absorption into one's inner self, which Pfleiderer thinks was Heraclitus' source of absolute knowledge, was the one thing he most despised.

Let us now consider the connection of Heraclitus with the Greek Mysteries, which Pfleiderer makes the basis of his interpretation of the whole philosophy. Pfleiderer has done a good work in emphasizing the religious character of the philosophy of the Ephesian. Lassalle and Teichmüller had already pointed it out. Failure to recognize this is the gravest fault in the critical work of Zeller. But as in Lassalle we found over-systemization of the logical idea, in Schuster of the empirical, in Teichmüller of the physical, so in Pfleiderer there is great over-systemization of the religious element. More strictly, it is a vast over-emphasis of one thought, namely, the indestructibility of life, or the alternating identity of life and death, which Pfleiderer claims to be a religious truth taken from the Mysteries, and out of which, as we have seen, he spins the whole philosophy of Heraclitus, including the doctrine of the eternal flux, the unity of opposites, and the fire. The slight grounds on which all this is based must have already impressed the reader with surprise that Pfleiderer should make so much out of it. The fact that Heraclitus lived in Ephesus and that Ephesus was a very religious city, is a fair specimen of the arguments by which he would establish a connection with the Mysteries. There have been preserved only three fragments in which Heraclitus makes any direct reference to the Greek Mysteries, all taken from Clement of Alexandria (Protrept. 2, pp. 19, 30; cp. frags. 124, 125, 127), and in these three passages other critics have found no sympathy with, but stern condemnation of the mystic cult. In the first passage where the νυκτιπόλοι, μάγοι, βάκχοι, λῆναι and μύσται are threatened with future fire, Pfleiderer admits condemnation of mystic abuses. But the third fragment, relating to the Dionysiac orgies, is the one upon which he most relies to establish the sympathy of our philosopher with the Mysteries. The passage is as follows: Εἰ μὴ γὰρ Διονύσῳ πομπὴν ἐποιεῦντο καὶ ὕμνεον ᾆσμα αἰδοίοισι, ἀναιδέστατα εἴργαστ᾽ ἄν· ὡυτὸς δὲ᾽ Ἀίδης καὶ Διόνυσος, ὅτεῳ μαίνονται καὶ ληναΐζουσι. "For were it not Dionysus to whom they institute a procession and sing songs in honor of the pudenda, it would be the most shameful action. But Dionysus, in whose honor they rave in bacchic frenzy, and Hades, are the same." Although this has usually been interpreted (by Schleiermacher, Lassalle, and Schuster) to mean that the excesses practiced in these ceremonies will be atoned for hereafter, since Dionysus under whose name they are carried on is identical with Pluto, the god of the lower world, Pfleiderer, interpreting it in a wholly different spirit, believes it to mean that these rites, although in themselves considered they would be most shameful, nevertheless have at least a partial justification from the fact that they are celebrated in honor of Dionysus, because since Dionysus and Pluto are the same, the rites are really a symbolism expressing the power of life over death and the indestructibility of life even in death. These vile phallus songs are in fact songs of triumph of life over death (Pfleiderer, p. 28). Although somewhat far-fetched, this is a possible interpretation of this obscure passage. This explanation is perhaps not more strained than the others that have been given (see below, frag. 127, crit. note). Granting it, and granting that Heraclitus here expresses a certain sympathy with, or at least does not express condemnation of the Mysteries, what follows? Surely, Pfleiderer would not seriously ask us to conclude from a single passage friendly to the religion of the Mysteries, that Heraclitus' whole philosophy or any part of it was drawn from them.

But this fragment has another and more important use for Pfleiderer. In the religious truth here expressed of the identity of Dionysus and Hades, that is, the identity of life and death, he finds the germ of all the Heraclitic philosophy. But the serious question immediately arises whether the philosophy of opposites grew out of this identity, or whether this identity was merely another illustration of the law of opposites. As Pfleiderer has produced no sufficient reason for believing differently, the natural conclusion is that, as elsewhere we find the unity of day and night, up and down, awake and asleep, so here we have the unity of the god of death and the god of life, as another illustration of the general law. To reverse this and say that in this particular antithesis we have the parent of all antitheses is very fanciful. Still further, we should infer from Pfleiderer's argument that the identity of Dionysus and Hades was a well known and accepted truth among the Mysteries, and that in the above fragment we find it in the very act of passing into the philosophy of the Ephesian. How much truth is there in this? So little that there is no record of the identity of these two gods before the time of Heraclitus. Later, to be sure, something of the kind appears. Dionysus represented at least five different gods, and in different times and places seems to have been identified with most of the principal deities. In Crete and at Delphi we hear of Zagreus, the winter Dionysus of the lower world. No doubt other instances might be shown where Dionysus was brought into some relation or other with a chthonian deity. But Heraclitus, if he had wished to develop a philosophy from the alternation of summer and winter and the mystic symbolism of life and death therein contained, would hardly have chosen so dubious an expression of it as the unity of Dionysus and Hades. We have no reason to regard this as anything else than one of the many paradoxical statements which he loved, of his law of opposites. Indeed, the genesis of this law is not so obscure that we need to force it out of a hidden mystic symbolism. Zeller in his introduction to Greek philosophy has well said that "philosophy did not need the myth of Kore and Demeter to make known the alternation of natural conditions, the passage from death to life and life to death; daily observation taught it" (Vol. 1, p. 60).

The intrinsic weakness of Pfleiderer's position is best seen when he attempts to pass to the doctrine of the flux. It taxes the imagination to see how the identity of life and death should lead to the universal principle $\pi\acute{a}\nu\tau a\ \chi\omega\rho\epsilon\tilde{\iota}\ \varkappa a\grave{\iota}\ o\grave{\upsilon}\delta\grave{\epsilon}\nu\ \mu\acute{\epsilon}\nu\epsilon\iota$. Pfleiderer would have us believe that the eternal flux was a subordinate thought—a mere picture to help the mind to conceive the primary metaphysical truth of the unity of opposites. We have already attempted to show that any explanation of the Heraclitic philosophy must be wrong which reduces the doctrine of the flux to a subordinate position. Here it is sufficient to add that if Heraclitus had been seeking a picture to illustrate the optimistic endurance of life even in death, and the rational unity and harmony of opposite powers, he could not possibly have chosen a more unfortunate figure than the ever-flowing river into which one cannot step twice. Pfleiderer, in saying that Heraclitus chose the picture of the evanescence of things to illustrate his law of opposites and the endurance of life, seems to have forgotten that on a previous page (above, p. 602) he said that the hopeless creed of the masses, against which the Ephesian was trying to establish the triumph of life, saw not too much permanence and constancy in the world, but too little.

We are forced, therefore, to conclude not only that Pfleiderer has failed to establish any especial dependence of Heraclitus upon the religion of the Greek Mysteries, but also that his supposed discovery that we have here a metaphysical philosophy developed from the material principle of the oscillating identity of life and death, is an assumption without basis in fact.

In redeeming the Ephesian from the charge of pessimism, Pfleiderer has done a good work. But here again he has gone too far, in finding not only a well grounded rational optimism in the doctrine of a world-ruling Order, but also a practical optimism in the idea of the *indestructibility of life*, an idea which, although it appears on every page of Pfleiderer's book, is not to be found in any saying of Heraclitus or in any record of his philosophy.

Section II.—Reconstructive.

I.

Having examined the four preceding fundamentally different views of the philosophy of Heraclitus, and having discovered that the opinions of modern critics on the tenor of this philosophy furnish a new and unexpected illustration of Heraclitus' own law of absolute instability, it remains to be considered whether it is possible to resolve, as he did, this general diversity into a higher unity, and in this case to verify his law that in all opposition there is harmony. If such a unity is sought as that attempted by Lassalle, Schuster, and Pfleiderer, it may be said at once that the task is impossible. All such ambitious attempts in constructive criticism in the case of Heraclitus are certain to result, as we have seen, in over-interpretation, and while they may leave a completed picture in the mind of the reader, they do not leave a true one. Not only is such a unified view of the philosophy of the Ephesian unattainable, but it is unnecessary. It is quite certain that had we before us his original book in its entirety, we should find therein no fully consistent system of philosophy. Yet it is just this fact that modern critics forget. While they point out errors and contradictions by the score in the books of their fellow critics, they allow for no inconsistencies on the part of the original philosopher. Presuppositions of harmony between all the sentences of an ancient writer have led to much violence of interpretation. Our interest in Heraclitus is not in his system as such, but in his great thoughts which have historic significance. These we should know, if possible, in their original meaning and in their connection with preceding and succeeding philosophy. Before concluding this introduction, then, it will be of advantage to recapitulate the results of the foregoing criticism, and to place together such conclusions concerning the chief Heraclitic thoughts as we have drawn either from the agreement or the disagreement of the various critics.

We shall best understand Heraclitus if we fix well in mind his immediate starting-point. As we found above in the examination of Pfleiderer's position (p. 47), the Ephesian philosopher was first and primarily a *preacher*. To him the people almost without exception, were blind, stupid, and beastly. Heraclitus hated them. They got no farther than crude sense perception (cp. frags. 4, 6, 3), failing not only to recognize the invisible harmony of the changing world, but even the change itself (cp. frag. 2). They believed things were fixed because they appeared so at first sight. They preferred the lower passions to the higher senses (cp. frag. 111). He is from first to last a misanthrope. He despises the people, yet as if constrained by a divine command, he must deliver his message (cp. frags. 1, 2). To understand Heraclitus we must free our minds from conceptions of every other Greek philosopher, except, perhaps, his fellow Ionians. Never afterwards did philosophy exhibit such seriousness. We can no more imagine Heraclitus at Athens than we can think of Socrates away from it. Although, as we shall see, the philosophy of Plato stood in vital connection with that of Heraclitus, no contrast could be greater than the half playful speculative style of the former, and the stern, oracular and dogmatic utterances of the latter. We shall find no parallel except in Jewish literature. Indeed, Heraclitus was a prophet. As the prophets of Israel *hurled* their messages in actual defiance at the people, hardly more does the Ephesian seem to care how his words are received, if only he gets them spoken. Not more bitter and misanthropic is Hosea in his denunciation of the people's sins (cp. ch. iv. 1, 2 ff.), than is our philosopher in his contempt for the stupidity and dullness of the masses. At the very opening of his book he says, from his lofty position of conscious superiority : " This Law which I unfold, men insensible and half asleep will not hear, and hearing, will not comprehend" (frag. 2 ; cp. frags. 3, 5, 94, 95).

Now what was the prime error of the people which so aroused the Ephesian, and what was the message which he had to deliver to them? Zeller is wrong in saying (Vol. 1, p. 576) that, according to Heraclitus, the radical error of the people was in attributing to things a permanence of being which they did not possess. In no passage does he censure the people for this. What he blames them for is their *insensibility*, for looking low when they ought to look high—in a word, for blindness to the Divine Law or the Universal Reason (frags. 2, 3, 4, 51, 45, 14). He blames them for not recognizing the beauty of strife (frag. 43), and the law of opposites (frag. 45). He blames them for their grossness and beastliness (frags. 86, 111). Finally, he blames them for their immorality (frag. 124), their silliness in praying to idols (frag. 126), and their imbecility in thinking they could purify themselves by sacrifices of blood (frag. 130). We see therefore how wholly impossible it is to understand Heraclitus unless we consider the ethical and religious character of his mind. Thus Zeller, in as far as he has attempted to give us a picture of Heraclitus' system, has failed by starting with the doctrine of the flux and overlooking the religious motive. This is not to say, as Pfleiderer has done, that the flux was merely a negative teaching. Next to the recognition of the Eternal Law, it was the most positive of his teachings, and was the ground of his influence upon subsequent thought. As such it is of chief interest to us; but as far as we wish to get a picture of Heraclitus' himself, we must think first of his religious and ethical point of departure. Thus the content of Heraclitus' message to his countrymen was *ethical*. It was a call to men everywhere to *wake up*, to purify their βαρβάρους ψυχάς, and see things in their reality.

What now was this reality which he with his finer insight saw, but which ruder souls were blind to? This brings us to the theoretical side or the philosophical content of Heraclitus' message. Here comes in the contribution of Teichmüller, who, as we saw, clearly pointed out that the great new thought of the Ephesian was the unity *in* the manifold, as opposed to the unity over against the manifold, taught by Xenophanes. It was the unity of opposition, the harmony of strife. It was Order immanent in ceaseless change. To use a phrase of Campbell's, "The Idea of the universe implies at once absolute activity and perfect law" (Plat. Theaet. Appendix, p. 244). This was the central thought of Heraclitus, "the grandeur of which," says Campbell, "was far beyond the comprehension of that time." But, it may be said, if we have rightly apprehended Heraclitus' position as a prophet and preacher, this was rather strong meat to feed the masses. But the πολλοί with Heraclitus was a very broad term. It included everybody. The arrogance of this man was sublime. Neither Homer nor

Hesiod nor Pythagoras nor Xenophanes escaped his lash (cp. frags. 16, 17, 119, 114). He had especially in mind the so-called "men of repute," and said they were makers and witnesses of lies (cp. frag. 118). The whole male population of Ephesus, he said, ought to be hung or expelled on account of their infatuation and blindness (cp. frag. 114). Addressing such an audience, indeed, his message had to be pitched high. We have in the Ephesian sage a man who openly claimed to have an insight superior to all the world, and the history of thought has vindicated his claim. Furthermore, it must be remembered that Heraclitus did, in a measure, try to make the world-ruling Law intelligible. He pictured it now as Justice, whose handmaids, the Erinyes, will not let the sun overstep his bounds (frag. 29); now as Fate, or the all-determining Destiny (Stobaeus, Ecl. i, 5, p. 178; cp. frag. 63); now as simple Law (frags. 23, 91), now as Wisdom (frag. 65), intelligent Will (frag. 19), God (frag. 36), Zeus (frag. 65). Respecting the latter term he expressly adds that it is misleading. So we see that Heraclitus did what some modern philosophers have been blamed for doing—he put his new thoughts into old religious formulas. But it was more justifiable in the case of the Ephesian. He did so, not to present a semblance of orthodoxy, but to try to make his idea intelligible. In fact, Heraclitus, no less than Xenophanes, was a fearless, outspoken enemy of the popular anthropomorphisms. "This world, the same for all," he says, "neither any of the gods nor any man has made, but it always was, and is, and shall be, an ever living fire, kindled and quenched according to law" (frag. 20; cp. frag. 126).

At this point it is natural to ask ourselves what,

more exactly considered, Heraclitus meánt by his Universal Order, his Divine Law, *κοινὸς λόγος*, etc. This inquiry fair criticism will probably not allow us to answer more concisely than has already been done. We have found ample reason for rejecting the notion that it was of a logical nature, or any objectification of that which is inherent in human thought. Yet it was not without human attributes. As fiery essence, it was identified with the universe and became almost material. As Order, it approached the idea of pure mathematical Relation or Form (cp. frag. 23, and Zeller, Vol. 1, p. 628, 3, and 620). As Wisdom, it was pictured as the intelligent power or efficient force that produces the Order. When we reflect what difficulty even at the present day we find in answering the simple question, What is Order? we are less surprised to find that the Ephesian philosopher did not always distinguish it from less difficult conceptions. We are, however, surprised and startled at the significance of the thought which this early Greek so nearly formulated, that the one permanent, abiding element in a universe of ceaseless change is mathematical relation. At any rate, while recognizing the want of perfect consistency and coördination in Heraclitus' system here, we shall be helped by keeping this in mind, that the system was pure pantheism. Too much stress cannot be laid upon Teichmüller's exposition of the history of the idea of Transcendent Reason, which first arose, not in Heraclitus, but in Anaxagoras. To the latter belongs the credit or the blame, whichever it may be, of taking the first step towards the doctrine of immateriality or pure spirit, which has influenced not only philosophy, but society to its foundations even to the present day. Heraclitus was guiltless of it. To him the world intelligence itself was a part of the world material—itself took part in the universal change.

In close connection with the Heraclitic Universal Order stands the doctrine of *strife* as the method of the evolution of the world, and the doctrine of the harmony and ultimately the unity of opposites—thoughts which were not only central in Heraclitus' system, but which, being too advanced for his time, have waited to be taken up in no small degree by modern science. It is unnecessary to repeat here the explanations of Schuster (above, pp. 15, 16), and particularly of Teichmüller (above, p. 27), which we found to indicate the correct interpretation of these thoughts. These principles are to Heraclitus the mediation between absolute change and perfect law. That which appears to the senses as rest and stability is merely the temporary equilibrium of opposite striving forces. It is harmony by tension (cp. frags. 45, 43, 46). This law, elementary in modern physics, is yet, as we shall presently see, not the whole content of the Heraclitic thought, although it is its chief import. But in the equilibrium of opposite forces we have at least relative rest, not motion. And of molecular motion Heraclitus knew nothing. How then did he conceive of apparent stability as absolute motion? This question supposes more exactness of thought than we look for in the Ephesian. The eternal flux was more generally conceived as absolute *perishability*. Nothing is permanently fixed. All is involved in the ceaseless round of life and death, growth and decay. Strictly, however, there is no contradiction here, since the rest of balanced forces is only relative rest. It is possibly not going too far to accept an illustration given by Ernst Laas (Idealismus u. Positivismus 1, p. 200) of Heraclitus' conception of absolute change under the dominion of law. He compares it to the actual path of our planets, which move neither in circles nor in exact ellipses, but under the influence of the attractive forces of moons and of other planets, or of comets, continually change both their course and their velocity, and yet all according to law.

In addition to the explanations now given, however, there is something more to be said concerning the unity or sameness of opposites. This teaching is very prominent in the Heraclitic fragments (cp. frags. 35, 36, 39, 43, 45, 46, 52, 57, 58, 59, 67, 78). This prominence was no doubt less in the original work, as the paradoxical character of these sayings has encouraged their preservation. But all the critics have failed to notice that we have in these fragments two distinct classes of oppositions which, though confused in Heraclitus' mind, led historically into different paths of development. The first is that unity of opposites which results from the fact that they are endlessly passing into one another. It must not be forgotten that this is a purely physical opposition, as has been pointed out by Zeller, Schuster and others, in refutation of the opinion of Lassalle, who fancied that he had found here a Hegelian logical identity of contradictories. As examples of this class of oppositions may be mentioned the identity of day and night (frag. 35), gods and men (frag. 67), alive and dead, asleep and awake (frag. 78). The identity of these oppositions means that they are not in themselves abiding conditions, but are continually and reciprocally passing into one another. As Heraclitus plainly says, they are the same because they are reciprocal transmutations of each other (frag. 78). But now we have another class of opposites to which this reasoning will not apply. "Good and evil," he says, "are the same" (frag. 57). This is simply that identity of opposites which developed into the Protagorean doctrine of *relativity*. The same thing may be good or evil according to the side from which you look at it. The passage from Hippolytus (Ref. haer. ix. 10 ; cp. frag. 52, sources) states the doctrine of relativity as plainly as it can be stated. " Pure and impure, he [Heraclitus] says, are one and the same, and drinkable and undrinkable are one and the same. 'Sea water,' he says, 'is very pure and very foul, for while to fishes it is drinkable and healthful, to men it is hurtful and unfit to drink.'" (Compare the opposition of just and unjust, frag. 61 ; young and old, frag. 78 ; beauty and ugliness, frag. 99 ; cp. frags. 104, 98, 60, 61, 51, 53.) This simple truth is so prominent in the Heraclitic sayings that we see how Schuster could have mistaken it for the whole content of the theory of opposites and ignored the more important doctrine of the other class. We see further that Plato's incorrect supposition that the Protagorean subjectivism was wholly an outgrowth of the Heraclitic flux, resulted from his insufficient acquaintance with the Ephesian's own writings. It was a characteristic of Heraclitus that, in a degree surpassing any other philosopher of antiquity, and comparable only to the discoveries of Greek mathematicians and of modern physical philosophers, he had an insight into truths beyond his contemporaries, but he knew not how to coördinate or use them. Having hit upon certain paradoxical relations of opposites, he hastened to group under his new law all sorts of oppositions. Some that cannot be included under either of the above classes appear in a passage from Aristotle (de Mundo, 5, p. 396 b 12; cp. frag. 59, sources; cp. Eth. Eud. vii. 1, p. 1235 a 26; frag. 43), where in the case of the opposites sharps and flats, male and female, the opposition becomes simple correlation and the unity, harmony.

The order of treatment brings us now to the Heraclitic flux, but we have been compelled so far to anticipate this in discussing the Universal Order and the Law of Opposites that but one or two points need be considered here. As we have seen in the study of Schuster and Teichmüller, the Heraclitic doctrine of the flux was a thoroughly radical one. Heaven and earth and all that they contain were caught in its fatal whirlpool. It exempted no immortal gods of the poets above us, no unchangeable realm of Platonic ideas around us, no fixed Aristotelian earth beneath us. It banished all permanence from the universe, and banished therewith all those last supports which men are accustomed to cling to. It introduced alarm into philosophy, and set men, even to the present day, asking, What can be saved from this general wreck? What is there absolutely permanent in the universe? This question, as we have seen, did not trouble Heraclitus himself, for, consistently or inconsistently, he had a foundation rock in his Universal Law, Reason or Order, which was his theoretical starting-point. Furthermore, concerning the flux, it is doubtful whether he ever pictured to himself such absolute instability as his words imply.

But we are tempted to ask, Is his system here really, as it first appears, inconsistent? Mr. Borden P. Bowne in his Metaphysics (p. 89) says that the Heraclitic theory of change thus extremely conceived " is intelligible and possible only because it is false."

Let us look at Mr. Bowne's argument. He has first shown in the same chapter that the Eleatic conception of rigid being without change is impossible, since in a world of absolute fixity, even the illusion of change would be impossible. Furthermore, he has shown that the vulgar conception of changeless being with changing states is untenable, since the "state of a thing expresses what the thing is at the time." Changing states would be uncaused and undetermined except as the being changes. There can be therefore no fixed useless core of being. In general there is no changeless being. All is change, all is becoming. Is there then, he asks, any permanence or identity whatever, or is the extreme Heraclitic position true? It is false. Why? Because, as in a world of Eleatic fixity, even the illusion of change would be impossible, so in a world of absolute change, even the appearance of rest would be impossible. There must be some abiding factor, that change may be known as change. There must be something permanent somewhere to make the notion of flow possible. This permanent something Mr. Bowne finds in the knowing subject— the conscious self. Having proceeded plainly up to this point, here he becomes mystical. The permanence of the conscious self, he continues, does not consist in any permanent substance of the soul. The soul forever changes equally with other being. The permanence consists in memory or self-consciousness. "How this is possible," he says, "there is no telling." The permanence and identity of the soul consists, however, only in its ability "to gather up its past and carry it with it."

In this argument, Mr. Bowne's first fallacy is in saying that in a world of absolute change there must be some permanent factor in order that the change itself may be known. This is meaningless. Permanent as regards what? Permanence as regards other moving factors is simply relative difference of change. Mr. Bowne seems to have committed the primitive error of supposing that because all things seem to move, he alone is fixed—like the earth in the Ptolemaic astronomy. According to his argument, if he were in a moving car and should meet another moving car, the perception of movement would be impossible. His reasoning assumes that by absolute change is meant uniform change all in one way, which would not be change at all, but absolute fixity. *Difference* is the essential element in change, and difference is all that is necessary to the idea of change. The assumption of permanent personality in order to make change itself possible is unnecessary. Mr. Bowne says that what constitutes permanence in the conscious self is its ability to gather up its past and carry it with it. But a stratifying rock or growing tree gathers up its past and carries it with it. But the apparent permanence in the case of the rock or tree is a temporarily abiding *form* or temporarily abiding spacial relations. The apparent permanence of personality may similarly consist wholly in a temporarily abiding form or relation, must in fact consist in this, since Mr. Bowne rejects any abiding soul substance. But temporarily abiding relations, the extreme Heracliteans do not deny, certainly not Heraclitus, to whom apparent rest was due to the temporary equilibrium of opposite balancing forces. We conclude, therefore, that Mr. Bowne's charge of falsity against the theory of the Heraclitic flux is not well substantiated. Here as ever we see the difference between modern and ancient philosophy. The former looks within, the latter without. Mr. Bowne seeks the abiding within himself. Heraclitus looked away from himself to the Universal Order without, which determined all things and himself.

But though the Heraclitic absolute flux is vindicated from objections of the above character, the question still remains unanswered whether the doctrine is consistent with his conception of absolute Order. Did not Heraclitus make the common mistake of hypostasizing law? Did he not conceive of law as something by which the action of things is predetermined, rather than as a mere abstraction from the action of things? No doubt he did even worse than this, for he ascribed to his $κοινὸς\ λόγος$, attributes which led Bernays and Teichmüller to believe that it was a self-conscious being, (a conclusion questioned by Zeller, Vol. 1, p. 609, 3). But yet again he saved his consistency here by identifying his Absolute with fire and thereby bringing it after all into the all-consuming vortex of endless change. But in the face of this all-embracing flux, the one idea which stands out most prominent in Heraclitus is the deep rationality of the world—the eternal Order. Nor in the last analysis are these two at variance, for any world must be rational to the beings in it, for the rationality of the world to us is only our adaptation to the world, which is involved in the very fact of our existence.

Concerning the cosmogony, it is worth while to recall the suggestive thought contained in the $χρησμοσύνη$ and $κόρος$ of Heraclitus. In our examination of Schuster's work we found reason to believe that the word $χρησμοσύνη$, which we may render "craving" or "longing," was used by the Ephesian to denote the character of the impulse or motive force by which the primitive world matter or fire evolved itself into the world of individual things. The records are too meagre to warrant much enlargement upon this idea; nevertheless it is important historically and in itself interesting. It is the beginning of that line of thought which finds the analogy to the original motive or creative power of the universe, not in man's intellectual but in his emotional nature, not in pure thought but in pure desire. It is opposed to the conception of Aristotle that the absolute first mover is pure intellect, the thought of thought (νόησις νοήσεως), and to the modern German enlargement of the same which began with the intellectual monads of Leibniz. On the other hand, it is in agreement with the idea brought out by Plato in his Symposium, the idea of Love as the source of development and immortality, and it reminds us later of Plotinus, who refuses to predicate thought or reason of the One but identifies it with the Good. The Heraclitic-Platonic notion is no less anthropomorphic than the Aristotelian-Leibnizian; but if the human mind must furnish forth some faculty to be singly hypostasized into God, we much prefer the richer emotional side to that of pure dry intellect or reason.

We come now to the Heraclitic ethics, the freshest and most vital part of his philosophy, but most misunderstood by all the critics. The practical ethical rule with Heraclitus is to follow the law of the state, which again is dependent upon the Divine Law (frags. 91, 100). From his standpoint this agrees with his injunction to live according to Nature (frag. 107). More broadly stated, men should follow the Universal as opposed to individual whims. "The Law of Reason is common, but the majority of people live as though they had an understanding of their own" (frag. 92). This leads us directly to the theoretical ethical principle which lay at the root of all Heraclitus' philosophy, and which we have outlined above (p. 58) in defining his starting point as that of a preacher and prophet. The highest good was *not* contentment (εὐαρέστησις), a statement taken from a single indefinite passage in Clement of Alexandria (Strom. ii. 21, p. 417 ; Clement is followed by Theodoretus, iv. p. 984, ed. Halle), and which, though adopted by Zeller, is as silly and impossible as the better authenticated statement that Heraclitus wept over everything. Such an ethical principle is at variance with every sentence of the Ephesian. He continually exhorts men, as we have seen, to arise, get out of their lethargy and wake up. His most pungent sarcasm is directed against the people who are in a state of indifference, sleepiness, contentment (frags. 2, 3, 5, 94, 95, etc.). The highest good with Heraclitus, therefore, is the greatest intellectually *activity*, the greatest *receptivity* to the divine reason around us, the greatest freedom from individual peculiarities and the greatest possession of that which is universal. "Human nature," he says, "does not possess understanding, but the divine does" (frag. 96). We must look away from ourselves to Nature around us. We must follow the universal Reason therein expressed. Proximately for men this is best found in the common, the normal, the customary, finally therefore in public law.

It will thus be noticed that we have in Heraclitus an emphatic expression of the type of ethics peculiar to the Greeks. Of the individual he thought little. "To me one is ten thousand if he be the best" (frag. 113). He blamed the Ephesians for their declaration of democracy (frag. 114). He would not have been able to appreciate those modern systems of ethics which make a moral law out of individual conscience and justify actions by good intentions. Heraclitus, as well as psychologists of recent times, seemed to appreciate the dangers of self-involution. His whole system is a protest against individual intensification. He will not have men roll themselves into a cocoon of a single system, or revolve in the circle of a single set of ideas. He will have them throw themselves open to the common light, keep every sense open and receptive to new impressions, and thereby attain truth, which is found in the universal alone.

The optimism which Pfleiderer justifies for Heraclitus does not stand in contradiction to the misanthropy that we have found to characterize him. His optimism was thoroughly Leibnizian. It was reasoned optimism, resulting in the strong conviction that the world is good, rational and orderly. Most men, to be sure, are fools, but it is their own fault, as they will not put themselves in right relation to the world. Gottlob Mayer, in a pamphlet entitled "Heraklit von Ephesus und Arthur Schopenhauer," has been at pains to prove that Heraclitus is a Schopenhauer pessimist. We cannot regard his attempt as successful. Our study of the Ephesian philosopher in the preceding pages has shown nothing more clearly than that the logical result of his metaphysics is not, as this author claims, pessimism, but quite the opposite. None of the passages which he cites (cp. frags. 86, 55, 84, 66, 20, 111) can be made to yield any pessimism beyond misanthropy, unless possibly the one from Lucian (Vit. Auct. c. 14,—$\Omega NHTH\Sigma$. τi $\gamma \dot{\alpha} \rho$ $\dot{o}$ $\alpha i \dot{\omega} \nu$ $\dot{\varepsilon} \sigma \tau \iota \nu$; HPA-$KAEITO\Sigma$. $\pi a \tilde{\iota} \varsigma$ $\pi a i \zeta \omega \nu$, $\pi \varepsilon \sigma \sigma \varepsilon \dot{\nu} \omega \nu$, $\delta \iota a \varphi \varepsilon \rho \dot{o} \mu \varepsilon \nu o \varsigma$, cp. frag.

79), where Time is compared to a child at play, now arranging, now scattering the pebbles. And yet nothing is conclusive from this. It refers evidently to the periodic creation and destruction of the world. Whether this world building is a pastime of Jove, or the product of fate or of love, makes no difference in this case, provided only the resulting world is one well disposed and rational.

II.

What has given rise to the reviving interest in Heraclitus attested by the monographs which have lately appeared? The modern world hardly hopes to get any new light from his oracular sayings gathered in mutilated fragments from Philo and Plutarch, from Clement and Origen. Such unhoped for light, however, as our introductory study has shown, may for some minds be found breaking in after all. But the interest in the philosopher of Ephesus is historical. The new discovery of the present half century is that the way to study philosophy is to study its history, and especially its genesis. The passion for origins has carried the interest back to Greek philosophy, and finally back to the beginnings of Greek philosophy. But there is still another reason for going back. In the confusion arising from the fall of the idealistic philosophy in Germany, it was first thought that it would be necessary to return to Kant and secure a new footing; not that any new light was seen emanating from Kant, but error having arisen, it was necessary to trace it to its source.

This movement has neither been successful nor does it promise to be. In fact, there is a certain weariness in philosophy of the whole modern subjective method.

The result has been that thinkers have turned away from it to the one objective side of modern philosophy, namely, the sciences. Those, however, who still retain their love of philosophy in its larger sense, are going back farther than Kant. They see that the whole Hume-Kantian-Fichtean movement was a digression; a sort of branch road, which to be sure had to be explored before philosophy could go on in safety, but which was found to lead nowhere in particular, and that, having thanked these investigators for their services rendered, we may decline to concern ourselves further with this digression, but go on with our search for objective results. In this search our starting point must be from that philosophy which is most free from this whole subjective tendency. Such is the philosophy of Greece. Considering therefore that the introspective method has not proved so fruitful as was hoped, and that it is at least more modest if not more rational to regard man as a part of Nature, rather than Nature as a part of man, students of philosophy are turning their attention to the Greek philosophers where the freer and more ingenuous conception rules..

(These two causes, therefore, the former, the passion for studying the origin and development of thought and the connection of different systems of thought, the latter, the need of disinfecting our minds from all the germs of a pathological introspective habit, and putting ourselves as an experiment in the position of those who took it for granted that Nature was larger than man, have led us back to Greek philosophy and especially to its sources.)

In either of these aspects Heraclitus is important. (He is a perfect, by all means the most perfect, illustration of those qualities which are usually supposed to characterize the Greek mind, namely, receptivity, unprejudiced freedom of thought, love of order, and trustful confidence in the unity of man and Nature. Of all the Greek schools these qualities were best represented by the Ionian thinkers who, coming before what has been called "the fall of man in Socrates," were free from the later dialectical disturbances. And of the Ionians, Heraclitus, the last, best incorporates them. But it is in the other aspect that the philosopher of Ephesus is most important, namely, in the origin and history of ideas. Let us notice summarily what has come from him.

To Heraclitus we trace the philosophy of change, prominent in subsequent Greek philosophy as γιγνόμενον, the indirect cause of the counter movement of Socrates and Plato with its powerful determining influences, central in modern times as *motion* in the philosophy of Hobbes and the ground principle in the important system of Trendelenburg, and finally in a logical transformation, prominent in both German and English thought as Werden or Becoming. To Heraclitus we trace the notion of Relativity, the central point in the doctrine of the Sophists, which by withdrawing every absolute standard of truth, threatened to destroy all knowledge and all faith, and which sent Socrates searching for something permanent and fixed in the concepts of the human mind, and so led to the finished results of Plato and Aristotle. To Heraclitus we trace some of the fundamental doctrines of the Stoics, namely, their abrogation of the antithesis of mind and matter and their return to pre-Socratic monism, their conception of Nature as larger than man and his complete subjection to it, and finally their doctrine of the future conflagration of the world, later an influential factor in Christianity.

These were the thoughts which were most important in their determining influence upon subsequent philosophy. The following, while in themselves no less important, were less directly involved in the history of opinion. Of these the first is the notion of Law and Order absolute and immanent in the world, an idea so large that no Greek follower could grasp it, and yet vital to Heraclitus' system, for without it his philosophy becomes the philosophy of desperation, the source among the ingenuous Greeks of the nihilism of Gorgias or the universal doubt of the skeptics, and among the brooding moderns the source of the pessimism of Schopenhauer. To Heraclitus again we trace, as Teichmüller has shown, the closely related doctrine of the immanence of God in the world, so that we have in him one source of the pantheistic systems. To Heraclitus, finally, we trace the physical law of opposites, the thought that all order and harmony and apparent permanence are the result of opposite tension, the balance of centrifugal and centripetal forces. Less involved in the history of philosophy, though most important to Heraclitus, and in themselves most interesting to us of modern times, are his great ethical thoughts which we have already outlined.

The determinative ideas of the Ephesian may be summed up in a word by saying that they represent all that way of thinking against which Socrates and Plato raised the whole weight of their authority. Without repeating here the facts, well enough known to everybody, of the Socratic reaction in Greek philosophy, we must sketch one or two phases of it in order to establish the influence and explain the final defeat of the Heraclitic philosophy. In Socrates, Plato, and Aristotle, philosophy underwent a change more radical than any other in its history, a change that was ultimately to revolutionize all thought, and through its influence on Christian theology, to enter as a large determining element into all western civilization. Heraclitus is the representative of what philosophy was before that change.

Socrates said he could not understand the book of Heraclitus. That was not strange. The Ephesian could have told him the reason why. The man who could learn nothing from the fields and trees (see Plato's Phaedrus, p. 230), who spent all his time in the Agora conversing with other men about virtue, and who never seemed to realize that there was a world above the heads and under the feet of men, was not likely to understand the book of Heraclitus. Could the Ephesian philosopher have taken the Athenian logician out and given him a few lessons from Nature at first hand, could he have induced him to desist for a while from his boring into human intellects in search of a definition, and got his gaze lifted up to the clouds and stars, and put him in actual contact with the περιέχον, he would have been an apter scholar with the book. But it is quite impossible even in fancy to think of these two men together. The communer with Nature, the stern misanthropic sage and prophet of Ephesus had no points in common with the society-loving Athenian sophist. They were radically different, and on this difference hangs the secret of the development of philosophy for two thousand years. Socrates was not a Greek at all. He denied the most characteristic traits of his nation. He was a modern in many true senses. He was a curiosity at Athens, and consequently very much in vogue.

Socrates represents the birth of self-consciousness. In

practicing his maieutic art to this end, he little thought that he was giving the death-blow to the most beautiful trait of his countrymen, namely, the instinctive, the unconscious, the naive. No doubt this new birth had to take place some time, but under Socrates' direction it was premature. The old methods were not yet dead. Here historians of philosophy err. They say the pre-Socratic philosophers of Nature had in vain tried to solve the problems of the world, and it was high time for a critical philosophy that should begin with man. In vain, indeed! Had the naturalists labored in vain when the foundation of the atomic philosophy had been laid in Abdera, that of mathematics in Italy, and a far-seeing metaphysics and ethics in Ephesus? Socrates and Plato took fright too easily at the Sophists. Their philosophy would have died with them. Not so that of Democritus, Pythagoras, and Heraclitus. Socrates was a professor of clear thinking. Clear thinking is in itself well, but two solid centuries of clear thinking from Descartes to Hegel have in modern times ended in failure. We long to know what *natural* thinking would have accomplished if it had been left an open field a while longer in Greece. Then again clear thinking was overdone. It was, to be sure, not Socrates' fault that his method was afterwards abused, but as a matter of fact it took in later history a pathological turn that has resulted in wide-spread evils. Over self-consciousness, too much inwardness and painful self-inspection, absence of trust in our instincts and of the healthful study of Nature, which in ethics are illustrated in modern questions of casuistry, and in philosophy in Cartesian doubt and the skepticism of Hume, characterize our worst faults. The philosophy and ethics of Heraclitus, as we have seen, stood in vital opposition to all these traits.

But there was another respect in which the fall of man took place in Socrates. The love of beauty and form, and particularly beauty of the human body, characterized all the Greeks until Socrates, but characterizes modern people in a relatively small degree. Socrates cared nothing for outward beauty, but to the surprise of his fellow-citizens laid all the emphasis upon moral beauty. (We will say he was too large hearted to have had a personal motive for so doing.) It may be that the Greeks estimated physical beauty relatively too high, but the rebound has been too great. Caught up by the genius of Plato and intensified by the tenor of his philosophy, and met six centuries later in Alexandria by a powerful current of the same tendency from Judea, it effected the complete destruction of the Greek idea, and with it of course of Greek art. In the medieval church, inherited moral deformity was a *sin* of such extreme import, that for it a man was to be forever damned; but inherited physical deformity was not only not a sin, but often a blessing, teaching him as it did the relative worthlessness of the earthly life and body. So far was the Greek idea reversed that the body, instead of being the type of beauty, became the type of impurity, and from being the support of the soul, became its contaminator. The "flesh," indeed, was the symbol of evil. The results in modern life are only too well known. (Among them may be mentioned the loss of appreciation of the worth of the present physical life in itself, failure to recognize the close connection of soul and body, and that the health of the former depends on the health of the latter, resulting in all the strange devices to secure the welfare of the soul in the face of persistent disregard of the laws of physical health, or in such attempts as that of sustaining the moral status of a community where all hygienic laws are violated. This idea has been ground into the popular mind by so long education that modern educators find it a serious problem how to correct it. (It is not merely physical education that is wanted, but a reconstruction of our notions about the relation of body and mind. The Socratic work must be in part undone, and we must get back more nearly to the pre-Socratic conception of *balance*, for to them physical ugliness was no less an evil than moral ugliness.

But there is still another aspect of the Socratic apostasy, as important as those we have mentioned, and so far-reaching in its effects that it determines modern thought even to the lowest ranks of society. In this movement begun by Socrates, but perfected by Plato and Aristotle, the central thought of the Heraclitic philosophy was denied, and denied with such power that now after twenty-two hundred years it hardly dares assert itself. We refer, of course, to the Platonic transcendentalism. It was designed to give the death-blow to Heraclitus, and it succeeded ultimately beyond the wildest hopes of its founders. Strictly it was begun by Anaxagoras. We have already seen with Teichmüller how the doctrine of transcendent reason gained its first characteristic, Pure Separation, in the Nous of Anaxagoras, its second, Identity, in the definitive work of Socrates. But it was Plato who elevated it into a great system and gave it to the world for a perpetual inheritance. Finally, Aristotle, as if the fates conspired to make this doctrine immortal, took it up and adapted it to unpoetical inductive minds. Heraclitus in a wonderful conception of the world had abolished every antithesis and enunciated a system of pure monism. The Socratic school reversed his plan and set up a dualism of universal and particular, noumenon and phenomenon, mind and body, spirit and matter, which has dominated all philosophy, religion and literature.

(It is with the origin of this dualism that we are concerned, not with the familiar history of its outcome, but yet we may recall what to the student of philosophy or even of history it is needless to more than mention, how this dualism fastened itself upon subsequent thought; how as realism and nominalism it divided the schoolmen; how as mind and matter it left Descartes in hopeless difficulty; how Spinoza founded a philosophy expressly to resolve it, but succeeded only by the artifice of terms; how Leibnitz solved the problem, though with too much violence, by use of the same boldness with which its founders established it; how Kant finally left the antithesis unexplained; how again as the material and immaterial it fixed itself in the psychology of Aristotle, who affirmed as the higher part of the human mind, the active Nous or principle of pure immateriality, cognizant of the highest things, identical with the divine Prime Mover, and immortal, thus constituting for man the highest glorification that he ever received from his own hand; how Thomas Aquinas, spokesman for a powerful church, adopted this psychology and fastened it upon the modern popular world; how finally, in the sphere of religion proper, the transcendentalism of Plato has grown into the belief in pure Spirit and spiritual existences, peopling heaven and earth, and holding communion with matter and body, though having absolutely nothing in common (if the paradox may be excused) with them. Such has been in part the wonderful expansion of the Platonic Idealism.

And what was all this for in the first place? It was raised primarily as a barrier against the dissolving power of the eternal flux of the Heracliteans. A philosophy had arisen in Greece that denied all permanence. Misunderstood by the Sophists and abused by Cratylus, it called out the protest of Socrates, at heart the sincerest man of his contemporaries. Man, impelled by that very faculty which connects him most closely with Nature, namely, the sense of dependence, demands something permanent and unchangeable, upon which he can base his laws, religion and philosophy. If he cannot find it in Nature or in Revelation, he will make it out of a part of himself. This is what Socrates and Plato did. Socrates, seeking the permanent for ethical motives, detesting Nature and failing to find there anything fixed and abiding, turned to man and man's manner of thinking. By analysis of thought he separated out general concepts which appeared to be the same for all. Plato, perhaps less in earnest than subsequent ages gave him credit for, hypostasized them, raised them into real *objective* existences, henceforth to become idols, convenient entities to fill all gaps in human reasoning, objects of the dreams of poets and the worship of the religious, archetypes from which a lazy philosophy could deduce the universe. How, we naturally ask, could this audacious piece of anthropomorphism, in which man deliberately took his own norms of thought, projected them outward, and elevated them into gods, impose itself upon the world as it did? There are two answers. First, it flattered men immensely, and like all anthropomorphisms, thereby won half the battle. Second, it did *not* succeed at once, but slumbered for four centuries, and finally, in the decadence of all systems of philosophy and the breaking up of the old civilization, awakened to supply the groundwork of a religious revival. Platonism fell dead on the Greek world. Plato, and Aristotle as well, shot over the heads of their fellows. The philosophy of the Academy was a brilliant piece of speculation such as only the age of Pericles could call out. After that, philosophy fell back into the old ways. The Older Academy dragged out a short existence and ·died. Zeno, a Cypriote, but in his desire for unity more Greek than Plato, studied first with Polemo, head of the Academy, but disappointed with Platonism, turned back to Heraclitus. His school, as well as the Epicureans and Skeptics, returned to the Heraclitic monism. These schools loyally upheld for three centuries the Greek idea of the unity of man and Nature. But philosophy itself was doomed and fated to pass over into religion on the one hand and mysticism on the other. Platonism was admirably adapted to this end. In luxurious Alexandria, the weary inductive method of Aristotle, which the Ptolemies had instituted in the Museum, soon yielded to the fascinating lazy philosophy of Plato. Philo the Jew, Plutarch the moralist, Valentinus the Gnostic, Origen the Christian, all yielded to it in greater or less degree. In Plotinus it reached its full fruitage. Porphyry, his pupil, relates that he was ashamed of having a body and was careless of its needs, so anxious was he ecstatically to absorb his soul in the Supra-rational Transcendent One. Here we have a last consequence of the Socratic doctrine of mind. Here we have the extreme opposition to the naturalism of Heraclitus which considered man as a subordinate part of Nature. Greek philosophy ended with the triumph of Socrates and the defeat of Heraclitus. The wealth of Plato and Aristotle was the bequest that was handed over to the coming centuries. The Greek naturalists were forgotten. It was reserved for the present century to revive and vindicate them.

In what has been said in setting in relief the philosophy of Heraclitus, it is obvious that we have been concerned with but two or three aspects of that of Socrates and Plato, namely, its transcendental, idealistic and subjective character. It is not necessary to add that were we referring to other sides of it, as for instance, the undeniable importance of Socrates' contribution to ethics, and that of Plato to ethics and religion as well as to real scientific thought, the result would be very different. And of the Idealism itself, its very fascination and prevalence argue that it meets some want of human beings. It is poetry, to be sure, but as poetry it has been and will still be useful in saving men from the dangers of coarse materialistic thought.

Heraclitus of Ephesus on Nature.

I.—It is wise for those who hear, not me, but the universal Reason, to confess that all things are one.[1]

II —To this universal Reason which I unfold, although it always exists, men make themselves insensible, both before they have heard it and when they have heard it for the first time. For notwithstanding that all things happen according to this Reason, men act as though they had never had any experience in regard to it when they attempt such words and works as I am now relating, describing each thing according to its nature and explaining how it is ordered. And some men are as ignorant of what

---

Sources.—I.—Hippolytus, Ref. haer. ix. 9. Context:—Heraclitus says that all things are one, divided undivided, created uncreated, mortal immortal, reason eternity, father son, God justice. "It is wise for those who hear, not me, but the universal Reason, to confess that all things are one." And since all do not comprehend this or acknowledge it, he reproves them somewhat as follows: "They do not understand how that which separates unites with itself; it is a harmony of oppositions like that of the bow and of the lyre" (=frag. 45)

Compare Philo, Leg. alleg. iii. 3, p. 88. Context, see frag 24.

II.—Hippolytus, Ref. haer. ix. 9. Context:—And that Reason always exists, being all and permeating all, he (Heraclitus) says in this manner. "To this universal," etc.

Aristotle, Rhet. iii. 5, p. 1407, b. 14. Context·—For it is very hard to punctuate Heraclitus' writings on account of its not being clear whether the words refer to those which precede or to those which follow. For instance, in the beginning of his work, where he says, "To Reason existing always men make themselves insensible." For here it is ambiguous to what "always" refers.

Sextus Empir. adv Math. vii. 132 —Clement of Alex. Stromata, v. 14, p. 716 —Amelius from Euseb. Praep. Evang xi. 19, p. 540.— Compare Philo, Quis. rer. div. haer. 43, p. 505.—Compare Ioannes Sicel. in Walz. Rhett. Gr. vi. p 95.

---

[1] The small figures in the translation refer to the critical notes, pp 115 ff

they do when awake as they are forgetful of what they do when asleep.²

III.—Those who hear and do not understand are like the deaf. Of them the proverb says: "Present, they are absent."

IV.—Eyes and ears are bad witnesses to men having rude souls.

V.—The majority of people have no understanding of the things with which they daily meet, nor, when instructed, do they have any right knowledge of them, although to themselves they seem to have.

VI.—They understand neither how to hear nor how to speak.

III.—Clement of Alex. Strom. v. 14, p. 718. Context :—And if you wish to trace out that saying, "He that hath ears to hear, let him hear," you will find it expressed by the Ephesian in this manner, "Those who hear," etc.
Theodoretus, Therap. i. p. 13, 49.

IV.—Sextus Emp. adv. Math. vii. 126. Context.—He (Heraclitus) casts discredit upon sense perception in the saying, "Eyes and ears are bad witnesses to men having rude souls" Which is equivalent to saying that it is the part of rude souls to trust to the irrational senses.
Stobaeus Floril. iv. 56.
Compare Diogenes Laert. ix. 7.

V.—Clement of Alex. Strom. ii. 2, p. 432.
M. Antoninus iv. 46. Context.—Be ever mindful of the Heraclitic saying that the death of earth is to become water, and the death of water is to become air, and of air, fire (see frag 25) And remember also him who is forgetful whither the way leads (comp. frag 73); and that men quarrel with that with which they are in most continual association (= frag. 93), namely, the Reason which governs all. And those things with which they meet daily seem to them strange; and that we ought not to act and speak as though we were asleep (= frag. 94), for even then we seem to act and speak.

VI.—Clement of Alex. Strom. ii. 5, p. 442. Context —Heraclitus, scolding some as unbelievers, says: "They understand neither how to hear nor to speak," prompted, I suppose, by Solomon, "If thou lovest to hear, thou shalt understand ; and if thou inclinest thine ear, thou shalt be wise."

VII.—If you do not hope, you will not win that which is not hoped for, since it is unattainable and inaccessible.

VIII.—Gold-seekers dig over much earth and find little gold.

IX.—Debate.

X.—Nature loves to conceal herself.

XI.—The God whose oracle is at Delphi neither speaks plainly nor conceals, but indicates by signs.

XII.—But the Sibyl with raging mouth uttering things solemn, rude and unadorned, reaches with her voice over a thousand years, because of the God.

VII.—Clement of Alex. Strom. ii. 4, p. 437. Context.—Therefore, that which was spoken by the prophet is shown to be wholly true, "Unless ye believe, neither shall ye understand." Paraphrasing this saying, Heraclitus of Ephesus said, "If you do not hope," etc.
Theodoretus, Therap i. p. 15, 51.

VIII.—Clement of Alex. Strom. iv. 2, p. 565.
Theodoretus, Therap i. p 15, 52

IX.—Suidas, under word ἀμφισβατεῖν 'Αμφισβατεῖν. ἔνιοι τὸ ἀμφισβητεῖν 'Ιωνες δὲ καὶ ἀγχιβατεῖν, καὶ ἀγχιβασιην 'Ηράκλειτος.

X.—Themistius, Or. v. p. 69 (= xii. p. 159). Context.—Nature according to Heraclitus, loves to conceal herself; and before nature the creator of nature, whom therefore we especially worship and adore because the knowledge of him is difficult.
Philo, Qu. in Gen. iv. 1, p. 237, Aucher.. Arbor est secundum Heraclitum natura nostra, quae se obducere atque abscondere amat.
Compare idem de Profug. 32, p. 573, de Somn. i. 2, p. 621; de Spec. legg. 8, p. 344.

XI—Plutarch, de Pyth orac. 21, p. 404. Context.—And I think you know the saying of Heraclitus that "The God," etc.
Iamblichus, de Myst. iii. 15.
Idem from Stobaeus Floril. lxxxi. 17.
Anon. from Stobaeus Floril. v. 72.
Compare Lucianus, Vit. auct. 14.

XII—Plutarch, de Pyth orac. 6, p. 397. Context.—But the Sibyl, with raging mouth, according to Heraclitus, uttering things solemn, rude and unadorned, reaches with her voice over a

ON NATURE. 87

XIII.—Whatever concerns seeing, hearing, and learning, I particularly honor.³ ✓

XIV.—Polybius iv. 40. Especially at the present time, when all places are accessible either by land or by water, we should not accept poets and mythologists as witnesses of things that are unknown, since for the most part they furnish us with unreliable testimony about disputed things, according to Heraclitus.

XV.—The eyes are more exact witnesses than the ears.⁴

---

thousand years, because of the God. And Pindar says that Cadmus heard from the God a kind of music neither pleasant nor soft nor melodious. For great holiness permits not the allurements of pleasures.

Clement of Alex. Strom. i. 15, p. 358.
Iamblichus, de Myst. iii. 8.
See also pseudo-Heraclitus, Epist. viii.

✓ XIII.—Hippolytus, Ref. haer. ix. 9, 10. Context —And that the hidden, the unseen and unknown to men is [better], he (Heraclitus) says in these words, "A hidden harmony is better than a visible" (= frag. 47). He thus praises and admires the unknown and unseen more than the known. And that that which is discoverable and visible to men is [better], he says in these words, "Whatever concerns seeing, hearing, and learning, I particularly honor," that is, the visible above the invisible. From such expressions it is easy to understand him. In the knowledge of the visible, he says, men allow themselves to be deceived as Homer was, who yet was wiser than all the Greeks; for some boys killing lice deceived him saying, "What we see and catch we leave behind; what we neither see nor catch we take with us" (frag. 1, Schuster). Thus Heraclitus honors in equal degree the seen and the unseen, as if the seen and unseen were confessedly one. For what does he say? "A hidden harmony is better than a visible," and, "Whatever concerns seeing, hearing, and learning, I particularly honor," having before particularly honored the invisible.

XV.—Polybius xii. 27. Context:—There are two organs given to us by nature, sight and hearing, sight being considerably the more truthful, according to Heraclitus, "For the eyes are more exact witnesses than the ears."

Compare Herodotus i. 8.

XVI.—Much learning does not teach one to have understanding, else it would have taught Hesiod and Pythagoras, and again Xenophanes and Hecataeus.

XVII.—Pythagoras, son of Mnesarchus, practised investigation most of all men, and having chosen out these treatises, he made a wisdom of his own—much learning and bad art.

XVIII.—Of all whose words I have heard, no one attains to this, to know that wisdom is apart from all.[5]

XIX.—There is one wisdom, to understand the intelligent will by which all things are governed through all.[6]

XX.—This world, the same for all, neither any of

XVI.—Diogenes Laert. ix. 1. Context.—He (Heraclitus) was proud and disdainful above all men, as indeed is clear from his work, in which he says, "Much learning does not teach," etc.

Aulus Gellius, N. A. praef. 12.

Clement of Alex. Strom. i. 19, p. 373.

Athenaeus xiii p. 610 B.

Iulianus, Or. vi. p. 187 D.

Proclus in Tim 31 F.

Serenus in Excerpt. Flor. Ioann. Damasc. ii. 116, p. 205, Meinek.

Compare pseudo-Democritus, fr. mor 140 Mullach.

XVII.—Diogenes Laert. viii. 6 Context.—Some say, foolishly, that Pythagoras did not leave behind a single writing. But Heraclitus, the physicist, in his croaking way says, "Pythagoras, son of Mnesarchus," etc

Compare Clement of Alex. Strom i. 21, p. 396.

XVIII.—Stobaeus Floril. iii 81.

XIX.—Diogenes Laert. ix. 1. Context:—See frag. 16.

Plutarch, de Iside 77, p. 382. Context:—Nature, who lives and sees, and has in herself the beginning of motion and a knowledge of the suitable and the foreign, in some way draws an emanation and a share from the intelligence by which the universe is governed, according to Heraclitus.

Compare Cleanthes H. in Iov. 36.

Compare pseudo-Linus, 13 Mullach.

XX —Clement of Alex. Strom. v. 14, p. 711. Context.—Heraclitus of Ephesus is very plainly of this opinion, since he recognizes the gods nor any man has made, but it always was, and is, and shall be, an ever living fire, kindled in due measure, and in due measure extinguished.[7]

XXI.—The transmutations of fire are, first, the sea; and of the sea, half is earth, and half the lightning flash.[8]

XXII.—All things are exchanged for fire and fire for all things, just as wares for gold and gold for wares.

---

that there is an everlasting world on the one hand and on the other a perishable, that is, in its arrangement, knowing that in a certain manner the one is not different from the other. But that he knew an everlasting world eternally of a certain kind in its whole essence, he makes plain, saying in this manner, "This world the same for all," etc.

Plutarch, de Anim. procreat. 5, p. 1014. Context:—This world, says Heraclitus, neither any god nor man has made; as if fearing that having denied a divine creation, we should suppose the creator of the world to have been some man.

Simplicius in Aristot. de cael. p. 132, Karst.
Olympiodorus in Plat. Phaed. p. 201, Finckh.
Compare Cleanthes H., Iov. 9.
Nicander, Alexiph. 174.
Epictetus from Stob. Floril. cviii. 60.
M. Antoninus vii. 9.
Just. Mart. Apol. p. 93 C.
Heraclitus, Alleg. Hom. 26.

XXI.—Clement of Alex. Strom. v. 14, p. 712. Context:—And that he (Heraclitus) taught that it was created and perishable is shown by the following, "The transmutations," etc.

Compare Hippolytus, Ref. haer. vi. 17.

XXII.—Plutarch, de EI. 8, p. 388. Context:—For how that (scil. first cause) forming the world from itself, again perfects itself from the world, Heraclitus declares as follows, "All things are exchanged for fire and fire for all things," etc.

Compare Philo, Leg. alleg. iii. 3, p. 89. Context, see frag. 24.
Idem, de Incorr. mundi 21, p. 508.—Lucianus, Vit. auct. 14.
Diogenes Laert. ix. 8.
Heraclitus, Alleg. Hom. 43.
Plotinus, Enn. iv. 8, p. 468.—Iamblichus from Stob. Ecl. 1. 41.
Eusebius, Praep. Evang. xiv. 3, p. 720.—Simplicius on Aristot. Phys. 6, a.

XXIII.—The sea is poured out and measured to the same proportion as existed before it became earth.⁹

XXIV.—Craving and Satiety.¹⁰

XXV.—Fire lives in the death of earth, air lives in the death of fire, water lives in the death of air, and earth in the death of water.¹¹

XXVI.—Fire coming upon all things, will sift and seize them.

XXIII.—Clement of Alex. Strom v 14, p. 712 (= Eusebius, P. E. xiii. 13, p. 676). Context.—For he (Heraclitus) says that fire is changed by the divine Reason which rules the universe, through air into moisture, which is as it were the seed of cosmic arrangement, and which he calls sea; and from this again arise the earth and the heavens and all they contain And how again they are restored and ignited, he shows plainly as follows, "The sea is poured out," etc.

XXIV.—Hippolytus, Ref. haer. ix 10. Context:—And he (Heraclitus) says also that this fire is intelligent and is the cause of the government of all things. And he calls it craving and satiety. And craving is, according to him, arrangement (διακόσμησις), and satiety is conflagration (ἐκπύρωσις). For, he says, "Fire coming upon all things will separate and seize them" (= frag. 26).

Philo, Leg. alleg. iii. 3, p. 88. Context.—And the other (scil. ὁ γονορρυής), supposing that all things are from the world and are changed back into the world, and thinking that nothing was made by God, being a champion of the Heraclitic doctrine, introduces craving and satiety and that all things are one and happen by change

Philo, de Victim. 6, p. 242.

Plutarch, de EI. 9, p. 389.

XXV.—Maximus Tyr. xli. 4, p. 489. Context.—You see the change of bodies and the alternation of origin, the way up and down, according to Heraclitus. And again he says, "Living in their death and dying in their life (see frag 67). Fire lives in the death of earth," etc.

M. Antoninus iv. 46. Context, see frag. 5.

Plutarch, de EI. 18, p. 392.

Idem, de Prim frig. 10, p. 949. Comp. pseudo-Linus 21, Mull.

XXVI.—Hippolytus, Ref. haer. ix 10. Context, see frag. 24.

Compare Aetna v. 536 · quod si quis lapidis miratur fusile robur, cogitet obscuri verissima dicta libelli, Heraclite, tui, nihil insuperabile ab igni, omnia quo rerum naturae semina iacta.

ON NATURE.   91

XXVII.—How can one escape that which never sets?[12]

XXVIII.—Lightning rules all.

XXIX.—The sun will not overstep his bounds, for if he does, the Erinyes, helpers of justice, will find him out.

XXX.—The limits of the evening and morning are the Bear, and opposite the Bear, the bounds of bright Zeus.

XXXI.—If there were no sun, it would be night.

XXVII.—Clement of Alex. Paedag. ii. 10, p. 229. Context:—For one may escape the sensible light, but the intellectual it is impossible to escape. Or, as Heraclitus says, "How can one escape that which never sets?"

XXVIII.—Hippolytus, Ref. haer. ix. 10. Context:—And he (Heraclitus) also says that a judgment of the world and all things in it takes place by fire, expressing it as follows, "Now lightning rules all," that is, guides it rightly, meaning by lightning, everlasting fire.

Compare Cleanthes H., Iovem 10.

XXIX.—Plutarch, de Exil. II, p. 604. Context:—Each of the planets, rolling in one sphere, as in an island, preserves its order. "For the sun," says Heraclitus, "will not overstep his bounds," etc.

Idem, de Iside 48, p. 370.

Comp. Hippolytus, Ref. haer. vi. 26.

Iamblichus, Protrept. 21, p. 132, Arcer.

Pseudo-Heraclitus, Epist. ix.

XXX.—Strabo i. 6, p. 3. Context:—And Heraclitus, better and more Homerically, naming in like manner the Bear instead of the northern circle, says, "The limits of the evening and morning are the Bear, and opposite the Bear, the bounds of bright Zeus." For the northern circle is the boundary of rising and setting, not the Bear.

XXXI.—Plutarch, Aq. et ign. comp. 7, p. 957.

Idem, de Fortuna 3, p. 98. Context:—And just as, if there were no sun, as far as regards the other stars, we should have night, as Heraclitus says, so as far as regards the senses, if man had not mind and reason, his life would not differ from that of the beasts.

Compare Clement of Alex. Protrept. II, p. 87.

Macrobius, Somn. Scip. i. 20.

XXXII.—The sun is new every day.

XXXIII.—Diogenes Laertius i. 23. He (scil. Thales) seems, according to some, to have been the first to study astronomy and to foretell the eclipses and motions of the sun, as Eudemus relates in his account of astronomical works. And for this reason he is honored by Xenophanes and Herodotus, and both Heraclitus and Democritus bear witness to him.

XXXIV.—Plutarch, Qu. Plat. viii. 4, p. 1007. Thus Time, having a necessary union and connection with heaven, is not simple motion, but, so to speak, motion in an order, having measured limits and periods. Of which the sun, being overseer and guardian to limit, direct, appoint and proclaim the changes and seasons which, according to Heraclitus, produce all things, is the helper of the leader and first God, not in small or trivial things, but in the greatest and most important.

XXXV.—Hesiod is a teacher of the masses. They suppose him to have possessed the greatest knowledge, who indeed did not know day and night. For they are one.[13]

XXXII.—Aristotle, Meteor. ii. 2, p 355 a 9. Context:—Concerning the sun this cannot happen, since, being nourished in the same manner, as they say, it is plain that the sun is not only, as Heraclitus says, new every day, but it is continually new.

Alexander Aphrod in Meteor. l. l. fol. 93 a.

Olympiodorus in Meteor. 1 l. fol. 30 a.

Plotinus, Enn. ii. 1, p. 97.

Proclus in Tim. p. 334 B.

Compare Plato, Rep vi. p 498 B

Olympiodorus in Plato, Phaed p. 201, Finckh.

XXXIV.—Compare Plutarch, de Def. orac. 12, p. 416.

M. Antoninus ix. 3.

Pseudo-Heraclitus, Epist. v

XXXV.—Hippolytus, Ref haer ix. 10. Context.—Heraclitus says that neither darkness nor light, neither evil nor good, are different, but they are one and the same. He found fault, therefore, with

ON NATURE.   93

XXXVI.—God is day and night, winter and summer, war and peace, plenty and want. But he is changed, just as when incense is mingled with incense, but named according to the pleasure of each.[14]

XXXVII.—Aristotle, de Sensu 5, p. 443 a 21. Some think that odor consists in smoky exhalation, common to earth and air, and that for smell all things are converted into this. And it was for this reason that Heraclitus thus said that if all existing things should become smoke, perception would be by the nostrils.

XXXVIII.—Souls smell in Hades.[15]

XXXIX.—Cold becomes warm, and warm, cold; wet becomes dry, and dry, wet.

XL.—It disperses and gathers, it comes and goes.[16]

Hesiod because he knew [not] day and night, for day and night, he says, are one, expressing it somewhat as follows: "Hesiod is a teacher of the masses," etc.

XXXVI.—Hippolytus, Ref. haer. ix. 10. Context:—For that the primal (Gr. πρῶτον, Bernays reads ποιητὸν, created) world is itself the demiurge and creator of itself, he (Heraclitus) says as follows: "God is day and," etc.

Compare idem, Ref. haer. v. 21.

Hippocrates, περὶ διαίτης i. 4, Littr.

XXXVIII.—Plutarch, de Fac. in orbe lun. 28, p. 943. Context:— Their (scil. the souls') appearance is like the sun's rays, and their spirits, which are raised aloft, as here, in the ether around the moon, are like fire, and from this they receive strength and power, as metals do by tempering. For that which is still scattered and diffuse is strengthened and becomes firm and transparent, so that it is nourished with the chance exhalation. And finely did Heraclitus say that "souls smell in Hades."

XXXIX.—Schol. Tzetzae, Exeget. Iliad. p. 126, Hermann. Context:—Of old, Heraclitus of Ephesus was noted for the obscurity of his sayings, "Cold becomes warm," etc.

Compare Hippocrates, περὶ διαίτης i. 21.

Pseudo-Heraclitus, Epist. v.—Apuleius, de Mundo 21.

XL.—Plutarch, de EI. 18, p. 392. Context, see frag. 41.

Compare pseudo-Heraclitus, Epist. vi.

XLI.—⟨Into the same river you could not step twice, for other ⟨and still other⟩ waters are flowing.

XLII.—†To those entering the same river, other and still other waters flow.†

XLIII.—Aristotle, Eth. Eud. vii. 1, p. 1235 a 26. And Heraclitus blamed the poet who said, " Would

---

XLI.—Plutarch, Qu nat. 2, p. 912. Context.—For the waters of fountains and rivers are fresh and new, for, as Heraclitus says, "Into the same river," etc.

Plato, Crat. 402 A. Context :—Heraclitus is supposed to say that all things are in motion and nothing at rest; he compares them to the stream of a river, and says that you cannot go into the same river twice (Jowett's transl.).

Aristotle, Metaph. iii. 5, p. 1010 a 13 Context:—From this assumption there grew up that extreme opinion of those just now mentioned, those, namely, who professed to follow Heraclitus, such as Cratylus held, who finally thought that nothing ought to be said, but merely moved his finger. And he blamed Heraclitus because he said you could not step twice into the same river, for he himself thought you could not do so once

Plutarch, de EI. 18, p. 392. Context ·—It is not possible to step twice into the same river, according to Heraclitus, nor twice to find a perishable substance in a fixed state, but by the sharpness and quickness of change, it disperses and gathers again, or rather not again nor a second time, but at the same time it forms and is dissolved, it comes and goes (see frag. 40).

Idem, de Sera num. vind. 15, p 559.

Simplicius in Aristot Phys. f. 17 a

XLII —Arius Didymus from Eusebius, Praep. evang. xv. 20, p. 821. Context.—Concerning the soul, Cleanthes, quoting the doctrine of Zeno in comparison with the other physicists, said that Zeno affirmed the perceptive soul to be an exhalation, just as Heraclitus did For, wishing to show that the vaporized souls are always of an intellectual nature, he compared them to a river, saying, "To those entering the same river, other and still other waters flow." And souls are exhalations from moisture Zeno, therefore, like Heraclitus, called the soul an exhalation.

Compare Sextus Emp Pyrrh. hyp. iii. 115.

XLIII —Plutarch, de Iside 48, p 370. Context —For Heraclitus in plain terms calls war the father and king and lord of all (=frag. 44), and he says that Homer, when he prayed—"Discord be damned

that strife were destroyed from among gods and men."
For there could be no harmony without sharps and
flats, nor living beings without male and female,
which are contraries.

XLIV.—War is the father and king of all, and has
produced some as gods and some as men, and has
made some slaves and some free.

XLV.—They do not understand how that which

---

from gods and human race," forgot that he called down curses on
the origin of all things, since they have their source in antipathy
and war.

Chalcidius in Tim. 295.
Simplicius in Aristot. Categ. p. 104 Δ, ed. Basil.
Schol. Ven. (A) ad Il. xviii, 107.
Eustathius ad Il. xviii. 107, p. 1113, 56.

XLIV.—Hippolytus, Ref. haer. ix. 9. Context —And that the father of all created things is created and uncreated, the made and the maker, we hear him (Heraclitus) saying, " War is the father and king of all," etc.

Plutarch, de Iside 48, p. 370. Context, see frag. 43.
Proclus in Tim. 54 A (comp. 24 B).
Compare Chrysippus from Philodem. π. εὐσεβείας, vii. p. 81, Gomperz.
Lucianus, Quomodo hist. conscrib. 2; Idem, Icaromen 8.

XLV.—Hippolytus, Ref. haer. ix. 9. Context, see frag 1.
Plato, Symp. 187 A. Context :—And one who pays the least attention will also perceive that in music there is the same reconciliation of opposites; and I suppose that this must have been the meaning of Heraclitus, though his words are not accurate; for he says that the One is united by disunion, like the harmony of the bow and the lyre (Jowett's transl.).

Idem, Soph. 242 D. Context :—Then there are Ionian, and in more recent times Sicilian muses, who have conceived the thought that to unite the two principles is safer; and they say that being is one and many, which are held together by enmity and friendship, ever parting, ever meeting (idem).

Plutarch, de Anim. procreat. 27, p. 1026. Context :—And many call this (scil. necessity) destiny. Empedocles calls it love and hatred; Heraclitus, the harmony of oppositions as of the bow and of the lyre.

Compare Synesius, de Insomn. 135 A
Parmenides v. 95, Stein.

separates unites with itself. It is a harmony of oppositions, as in the case of the bow and of the lyre.[17]

XLVI.—Aristotle, Eth. Nic. viii. 2, p. 1155 b 1. In reference to these things, some seek for deeper principles and more in accordance with nature. Euripides says, "The parched earth loves the rain, and the high heaven, with moisture laden, loves earthward to fall." And Heraclitus says, "The unlike is joined together, and from differences results the most beautiful harmony, and all things take place by strife."

XLVII.—The hidden harmony is better than the visible.[18 and 3]

XLVIII.—Let us not draw conclusions rashly about the greatest things.

XLIX.—Philosophers must be learned in very many things.

L.—The straight and crooked way of the wool-carders is one and the same.[19]

XLVI.—Compare Theophrastus, Metaph. 15.
Philo, Qu. in Gen. iii. 5, p. 178, Aucher.
Idem, de Agricult. 31, p. 321.
XLVII.—Hippolytus, Ref. haer. ix. 9-10. Context, see frag. 13.
Plutarch, de Anim. procreat 27, p. 1026. Context.—Of the soul nothing is pure and unmixed nor remains apart from the rest, for, according to Heraclitus, "The hidden harmony is better than the visible," in which the blending deity has hidden and sunk variations and differences.
Compare Plotinus, Enn. i. 6, p. 53.
Proclus in Cratyl. p. 107, ed. Boissonad.
XLVIII.—Diogenes Laert. ix. 73. Context·—Moreover, Heraclitus says, "Let us not draw conclusions rashly about the greatest things." And Hippocrates delivered his opinions doubtfully and moderately
XLIX.—Clement of Alex. Strom v. 14, p. 733. Context:—Philosophers must be learned in very many things, according to Heraclitus. And, indeed, it is necessary that "he who wishes to be good shall often err."
L.—Hippolytus, Ref. haer. ix. 10. Context:—And both straight

ON NATURE.   97

LI.—Asses would choose stubble rather than gold.

LII.—Sea water is very pure and very foul, for, while to fishes it is drinkable and healthful, to men it is hurtful and unfit to drink.

LIII.—Columella, de Re Rustica viii. 4. Dry dust and ashes must be placed near the wall where the roof or eaves shelter the court, in order that there may be a place where the birds may sprinkle themselves, for with these things they improve their wings and feathers, if we may believe Heraclitus, the Ephesian, who says, "Hogs wash themselves in mud and doves in dust."

LIV.—They revel in dirt.

---

and crooked, he (Heraclitus) says, are the same: "The way of the wool-carders is straight and crooked." The revolution of the instrument in a carder's shop (Gr. γναφείῳ Bernays, γραφείῳ vulg.) called a screw is straight and crooked, for it moves at the same time forward and in a circle. "It is one and the same," he says.

Compare Apuleius, de Mundo 21.

LI.—Aristotle, Eth. Nic. x. 5, p. 1176 a 6. Context.—The pleasures of a horse, a dog, or a man, are all different. As Heraclitus says, "Asses would choose stubble rather than gold," for to them there is more pleasure in fodder than in gold.

LII.—Hippolytus, Ref. haer. ix. 10. Context:—And foul and fresh, he (Heraclitus) says, are one and the same. And drinkable and undrinkable are one and the same. "Sea water," he says, "is very pure and very foul," etc.

Compare Sextus Empir. Pyrrh. hyp. i. 55.

LIII.—Compare Galenus, Protrept. 13, p. 5, ed. Bas.

LIV.—Athenaeus v. p. 178 F. Context:—For it would be unbecoming, says Aristotle, to go to a banquet covered with sweat and dust. For a well-bred man should not be squalid nor slovenly nor delight in dirt, as Heraclitus says.

Clement of Alex. Protrept. 10, p. 75.

Idem, Strom. i. 1, p. 317; ii. 15, p. 465.

Compare Sextus Empir. Pyrrh. hyp. i. 55.

Plotinus, Enn. i. 6, p. 55.

Vincentius Bellovac. Spec. mor. iii. 9, 3.

LV.—Every animal is driven by blows.[20]

LVI.—The harmony of the world is a harmony of oppositions, as in the case of the bow and of the lyre.[21]

LVII.—Good and evil are the same.

LVIII.—Hippolytus, Ref. haer. ix. 10. And good and evil (scil. are one). The physicians, therefore, says Heraclitus, cutting, cauterizing, and in every way torturing the sick, complain that the patients do not pay them fitting reward for thus effecting these benefits— †and sufferings.†

LV.—Aristotle, de Mundo 6, p. 401 a 8 (= Apuleius, de Mundo 36; Stobaeus, Ecl. 1. 2, p. 86). Context —Both wild and domestic animals, and those living upon land or in air or water, are born, live and die in conformity with the laws of God. "For every animal," as Heraclitus says, "is driven by blows" (πληγῇ Stobaeus cod. A, Bergkius et al.; vulg. τὴν γῆν νέμεται, every animal feeds upon the earth)

LVI.—Plutarch, de Tranquill. 15, p. 473. Context —"For the harmony of the world is a harmony of oppositions (Gr. παλίντονος ἁρμονίη, see Crit. Note 21), as in the case of the bow and of the lyre. And in human things there is nothing that is pure and unmixed. But just as in music, some notes are flat and some sharp, etc.

Idem, de Iside 45, p 369. Context.—"For the harmony of the world is a harmony of opposition, as in the case of the bow and of the lyre," according to Heraclitus; and according to Euripides, neither good nor bad may be found apart, but are mingled together for the sake of greater beauty.

Porphyrius, de Antro. nymph. 29.

Simplicius in Phys fol. 11 a.

Compare Philo, Qu. in Gen. iii. 5, p 178, Aucher.

LVII.—Hippólytus, Ref. haer. ix. 10. Context, see frag. 58.

Simplicius in Phys. fol. 18 a. Context :—All things are with others identical, and the saying of Heraclitus is true that the good and the evil are the same.

Idem on Phys. fol. 11 a

Aristotle, Top. viii. 5, p 159 b 30.

Idem. Phys. i. 2, p. 185 b 20.

LVIII.—Compare Xenophon, Mem. i. 2, 54.

Plato, Gorg. 521 E; Polit. 293 B.

Simplicius in Epictetus 13, p. 83 D and 27, p. 178 A, ed. Heins.

LIX.—Unite whole and part, agreement and disagreement, accordant and discordant; from all comes one, and from one all.

LX.—They would not know the name of justice, were it not for these things.[22]

LXI.—Schol. B. in Iliad iv. 4, p. 120 Bekk. They say that it is unfitting that the sight of wars should please the gods. But it is not so. For noble works delight them, and while wars and battles seem to us terrible, to God they do not seem so. For God in his dispensation of all events, perfects them into a harmony of the whole, just as, indeed, Heraclitus says that to God all things are beautiful and good and right, though men suppose that some are right and others wrong.

LXII.—We must know that war is universal and strife right, and that by strife all things arise and †are used.†[23]

LIX.—Aristotle, de Mundo 5, p. 396 b 12 (= Apuleius, de Mundo 20; Stobaeus, Ecl. i. 34, p. 690). Context:—And again art, imitator of nature, appears to do the same. For in painting, it is by the mixing of colors, as white and black or yellow and red, that representations are made corresponding with the natural types. In music also, from the union of sharps and flats comes a final harmony, and in grammar, the whole art depends on the blending of mutes and vocables. And it was the same thing which the obscure Heraclitus meant when he said, "Unite whole and part," etc.

Compare Apuleius, de Mundo 21.

Hippocrates π. τροφῆς 40; π. διαίτης i.

LX.—Clement of Alex. Strom. iv. 3, p. 568. Context.—For the Scripture says, the law is not made for the just man. And Heraclitus well says, "They would not know the name of justice, were it not for these things."

Compare pseudo-Heraclitus, Epist. vii.

LXI.—Compare Hippocrates, περὶ διαίτης i. 11.

LXII.—Origen, cont. Celsus vi. 42, p. 312 (Celsus speaking). Context:—There was an obscure saying of the ancients that war was divine, Heraclitus writing thus, "We must know that war," etc.

Compare Plutarch, de Sol. animal. 7, p. 964.

Diogenes Laert. ix. 8.

LXIII.—For it is wholly destined———.

LXIV.—Death is what we see waking. What we see in sleep is a dream.[24]

LXV.—There is only one supreme Wisdom. It wills and wills not to be called by the name of Zeus.[25]

LXVI.—The name of the bow is life, but its work is death.

LXVII.—Immortals are mortal, mortals immortal, living in their death and dying in their life.

LXIII.—Stobaeus Ecl. i. 5, p. 178. Context.—Heraclitus declares that destiny is the all-pervading law. And this is the etherial body, the seed of the origin of all things, and the measure of the appointed course. All things are by fate, and this is the same as necessity. Thus he writes, "For it is wholly destined———" (The rest is wanting).

LXIV.—Clement of Alex. Strom. iii. 3, p. 520 Context.—And does not Heraclitus call death birth, similarly with Pythagoras and with Socrates in the Gorgias, when he says, "Death is what we see waking. What we see in sleep is a dream"?

Compare idem v. 14, p 712. Philo, de Ioseph. 22, p. 59.

LXV.—Clement of Alex. Strom. v. 14, p. 718 (Euseb. P. E. xiii. 13, p. 681) Context.—I know that Plato also bears witness to Heraclitus' writing, "There is only one supreme Wisdom. It wills and wills not to be called by the name of Zeus." And again, "Law is to obey the will of one" (= frag. 110).

LXVI.—Schol. in Iliad i. 49, fr. Cramer, A P. iii. p. 122. Context.—For it seems that by the ancients the bow and life were synonymously called βιός. So Heraclitus, the obscure, said, "The name of the bow is life, but its work is death."

Etym. magn under word βιός

Tzetze's Exeg. in Iliad, p. 101 Herm.

Eustathius in Iliad i. 49, p. 41

Compare Hippocrates, π. τροφῆς 21.

LXVII.—Hippolytus, Ref. haer. ix 10. Context:—And confessedly he (Heraclitus) asserts that the immortal is mortal and the mortal immortal, in such words as these, "Immortals are mortal," etc.

Numenius from Porphyr. de Antro nymph. 10. Context, see frag 72.

LXVIII.—To souls it is death to become water, and to water it is death to become earth, but from earth comes water, and from water, soul.

LXIX.—The way upward and downward are one and the same.

Philo, Leg. alleg. i. 33, p. 65.
Idem, Qu. in Gen. iv. 152, p. 360 Aucher.
Maximus Tyr. x. 4, p. 107. Idem, xli. 4, p. 489.
Clement of Alex. Paed. iii. 1, p. 251.
Hierocles in Aur. carm. 24.
Heraclitus, Alleg. Hom. 24, p. 51 Mehler.
Compare Lucianus, Vit. auct. 14.
Dio Cassius frr. i—xxxv. c. 30, t. i. p. 40 Dind.
Hermes from Stob. Ecl. i. 39, p. 768. Idem, Poemand. 12, p. 100.

LXVIII.—Clement of Alex. Strom. vi. 2, p. 746. Context.—(On plagiarisms) And Orpheus having written, "Water is death to the soul and soul the change from water; from water is earth and from earth again water, and from this the soul welling up through the whole ether"; Heraclitus, combining these expressions, writes as follows: "To souls it is death," etc.

Hippolytus, Ref. haer. v. 16. Context:—And not only do the poets say this, but already also the wisest of the Greeks, of whom Heraclitus was one, who said, "For the soul it is death to become water."

Philo, de Incorr. mundi 21, p. 509. Proclus in Tim. p. 36 C.
Aristides, Quintil. ii. p. 106, Meib.
Iulianus, Or. v. p 165 D.
Olympiodorus in Plato, Gorg. p. 357 Iahn; Idem, p 542

LXIX.—Hippolytus, Ref. haer. ix. 10. Context:—Up and down he (Heraclitus) says are one and the same. "The way upward and downward are one and the same."

Diogenes Laert. ix. 8. Context:—Heraclitus says that change is the road leading upward and downward, and that the whole world exists according to it.

Cleomedes, π. μετεώρων i. p. 75, Bak.
Maximus Tyr. xli. 4, p. 489.
Plotinus, Enn. iv. 8, p. 468.
Tertullian, adv. Marc. ii. 28.
Iamblichus from Stob. Ecl. i. 41.
Compare Hippocrates, π. τροφῆς 45.
M. Antoninus vi. 17.

LXX.—The beginning and end are common.

LXXI. The limits of the soul you would not find out, though you should traverse every way.

LXXII.—To souls it is joy to become wet.[26]

LXXIII.—A man when he is drunken is led by a beardless youth, stumbling, ignorant where he is going, having a wet soul.

LXXIV.—The dry soul is the wisest and best.[27]

Philo, de Incorr. mundi 21, p. 508.
Idem, de Somn. i. 24, p. 644.
Idem, de vit. Moys. i. 6, p. 85.
Musonius from Stob. Flo. 108, 60.

LXX.—Porphyry from Schol. B. Iliad xiv. 200, p. 392, Bekk. Context:—For the beginning and end on the periphery of the circle are common, according to Heraclitus.
Compare Hippocrates, π. τόπων τῶν κατ' ἄνθρωπον, 1.
Idem, π. διαίτης i. 19; π. τροφῆς, 9.
Philo, Leg. alleg. i. 3, p. 44. Plutarch, de EI. 8, p. 388.

LXXI.—Diogenes Laert. ix. 7. Context:—And he (Heraclitus) also says, "The limits of the soul you would not find out though you traverse every way," so deep lies its principle (οὕτω βαθὺν λόγον ἔχει).
Tertullian, de Anima 2.
Compare Hippolytus, Ref. haer. v. 7.
Sextus, Enchir. 386.

LXXII.—Numenius from Porphyry, de Antro nymph. 10. Context:—Wherefore Heraclitus says: To souls it is joy, not death, to become wet. And elsewhere he says: We live in their death and they live in our death (frag. 67).

LXXIII.—Stobaeus Floril. v. 120.
Compare M. Antoninus iv. 46. Context, see frag. 5.

LXXIV.—Plutarch, Romulus 28. Context:—For the dry soul is the wisest and best, according to Heraclitus. It flashes through the body as the lightning through the cloud (= fr. 63, Schleiermacher).
Aristides, Quintil. ii. p. 106.
Porphyry, de Antro nymph. 11.
Synesius, de Insomn. p. 140 A Petav.
Stobaeus Floril. v. 120.
Glycas, Ann. i. p. 74 B (compare 116 A).
Compare Clement of Alex. Paedag. ii. 2, p. 184.
Eustathius in Iliad xxiii. 261, p. 1299, 17 ed. Rom.

ON NATURE. 103

LXXV.—†The dry beam is the wisest and best soul.†

LXXVI.—†Where the land is dry, the soul is wisest and best.†[27]

LXXVII.—Man, as a light at night, is lighted and extinguished.[28]

· LXXVIII.—Plutarch, Consol. ad Apoll. 10, p. 106. For when is death not present with us? As indeed Heraclitus says: Living and dead, awake and asleep, young and old, are the same. For these several states are transmutations of each other.

LXXIX.—Time is a child playing at draughts, a child's kingdom.

LXXV.—Philo from Euseb. P. E. viii. 14, p. 399.
Musonius from Stob. Floril. xvii. 43.
Plutarch, de Esu. carn. i. 6, p. 995.
Idem, de Def. orac. 41, p. 432.
Galenus, π. τῶν τῆς ψυχῆς ἠθῶν 5, t. i. p. 346, ed. Bas.
Hermeias in Plat. Phaedr. p. 73, Ast.
Compare Porphyry, ἀφορμ. πρὸς τὰ νοητά 33, p. 78 Holst.; Ficinus, de Immort. anim. viii. 13.
LXXVI.—Philo from Euseb. P. E. vi. 14, p. 399.
Idem, de Provid. ii. 109, p. 117, Aucher.
LXXVII.—Clement of Alex. Strom. iv. 22, p. 628. Context:—Whatever they say of sleep, the same must be understood of death, for it is plain that each of them is a departure from life, the one less, the other more. Which is also to be received from Heraclitus: Man is kindled as a light at night; in like manner, dying, he is extinguished. And living, he borders upon death while asleep, and, extinguishing sight, he borders upon sleep when awake.
Compare Sextus Empir. adv. Math. vii. 130.
Seneca, Epist. 54.
LXXVIII.—Compare Plutarch, de EI. 18, p. 392.
Clement of Alex. Strom. iv. 22, p. 628. Context, see frag. 77.
Sextus Empir. Pyrrh. hyp. iii. 230.
Tzetze's Chil. ii. 722.
LXXIX.—Hippolytus, Ref. haer. ix. 9.
Proclus in Tim. 101 F. Context:—And some, as for example Heraclitus, say that the creator in creating the world is at play.
Lucianus, Vit. auct. 14. Context:—And what is time? A child at play, now arranging his pebbles, now scattering them.

LXXX.—I have inquired of myself.[29]

LXXXI.—Into the same river we both step and do not step. We both are and are not.

LXXXII.—It is weariness upon the same things to labor and by them to be controlled.[30]

Clement of Alex. Paedag. i. 5, p. 111.
Iamblichus from Stob. Ecl. ii 1, p 12.
Compare Plato, Legg. x. 903 D. Philo, de vit. Moys. i. 6, p. 85.
Plutarch, de EI. 21, p 393.
Gregory Naz Carm ii 85, p. 978 ed. Bened.

LXXX.—Diogenes Laert. ix. 5. Context —And he (Heraclitus) was a pupil of no one, but he said he inquired of himself and learned everything by himself.

Plutarch, adv. Colot. 20, p. 1118. Context.—And Heraclitus, as though he had been engaged in some great and solemn task, said, "I have been seeking myself." And of the sentences at Delphi, he thought the "Know thyself" to be the most divine.

Dio Chrysost. Or. 55, p. 282, Reiske.
Plotinus, Enn. iv. 8, p. 468.
Tatianus, Or. ad Graec. 3.
Iulianus, Or. vi. p 185 A.
Proclus in Tim 106 E
Suidas, under word Ποστοῦμος.
Compare Philo, de Ioseph. 22, p. 59.
Clement of Alex. Strom. ii. 1, p. 429.
Plotinus, Enn. v. 9, p. 559.

LXXXI —Heraclitus, Alleg. Hom 24
Seneca, Epist. 58. Context —And I, while I say these things are changed, am myself changed. This is what Heraclitus means when he says, into the same river we descend twice and do not descend, for the name of the river remains the same, but the water has flowed on. This in the case of the river is more evident than in case of man, but none the less does the swift course carry us on.

Compare Epicharmus, fr. B 40, Lorenz.
Parmenides v. 58, Stein.

LXXXII —Plotinus, Enn. iv. 8, p. 468.
Iamblichus from Stob. Ecl. 1. 41, p. 906. Context —For Heraclitus assumed necessary changes from opposites, and supposed that souls traversed the way upward and downward, and that to continue in the same condition is weariness, but that change brings rest (= fr. 83).

ON NATURE. 105

LXXXIII.—In change is rest.

LXXXIV.—A mixture separates when not kept in motion.

LXXXV.—Corpses are more worthless than excrement.

LXXXVI.—Being born, they will only to live and die, or rather to find rest, and they leave children who likewise are to die.

LXXXVII.—Plutarch, de Orac. def. 11, p. 415.

Aeneas, Gaz. Theophrast. p. 9.
Compare Hippocrates, π. διαίτης i. 15.
Philo, de Cherub. 26, p. 155.
LXXXIII.—Plotinus, Enn. iv. 8, p. 468.
Idem, iv. 8, p. 473.
Iamblichus from Stob. Ecl. i. 41, p. 906. Context, see frag 82.
Idem, p. 894.
Aeneas, Gaz. Theophrast. p. 9, Barth.
Idem, p. 11.
LXXXIV.—Theophrastus, de Vertigine 9, p. 138 Wimmer.
Alexander Aprod. Probl. p. 11, Usener. Context.—A mixture (ὁ κυκεών), as Heraclitus says, separates unless some one stirs it.
Compare Lucian, Vit. auct. 14.
M. Antoninus iv. 27.
LXXXV.—Strabo xvi. 26, p. 784. Context.—They consider dead bodies equal to excrement, just as Heraclitus says, "Corpses are more worthless," etc.
Plutarch, Qu. conviv. iv. 4, p. 669.
Pollux, Onom. v. 163.
Origen, c. Cels. v. 14, p 247.
Julian, Or. vii. p. 226 C.
Compare Philo, de Profug. ii. p. 555.
Plotinus, Enn. v. 1, p. 483.
Schol. V. ad Iliad xxiv. 54, p. 630, Bekk.
Epictetus, Diss. ii. 4, 5.
LXXXVI.—Clement of Alex. Strom. iii. 3, p. 516. Context:—Heraclitus appears to be speaking evil of birth when he says, "Being born, they wish only to live," etc.
LXXXVII.—The reference is to the following passage from Hesiod:

Those who adopt the reading ἡβῶντος (*i. e.* at man's estate, see Hesiod, fr. 163, ed. Goettling) reckon a generation at thirty years, according to Heraclitus, in which time a father may have a son who is himself at the age of puberty.

LXXXVIII.—Io. Lydus de Mensibus iii. 10, p. 37, ed. Bonn. Thirty is the most natural number, for it bears the same relation to tens as three to units. Then again it is the monthly cycle, and is composed of the four numbers 1, 4, 9, 16, which are the squares of the units in order. Not without reason, therefore, does Heraclitus call the month a generation.

LXXXIX.—In thirty years a man may become a grandfather.

XC.—M. Antoninus vi. 42. We all work together to one end, some consciously and with purpose, others unconsciously. Just as indeed Heraclitus, I think, says that the sleeping are co-workers and fabricators of the things that happen in the world.[31]

XCI.—The Law of Understanding is common to all. Those who speak with intelligence must hold fast to that which is common to all, even more strongly than

---

ἐννέα τοι ζώει γενεὰς λακέρυζα κορώνη
ἀνδρῶν ἡβώντων · ἔλαφος δέ τε τετρακόρων ὅς
τρεῖς δ᾿ ἐλάφους ὁ κόραξ γηράσκεται. αὐτὰρ ὁ φοίνιξ
ἐννέα τοὺς κόρακας · δέκα δ᾿ ἡμεῖς τοὺς φοίνικας
νύμφαι ἐυπλόκαμοι, κοῦραι Διὸς αἰγιόχοιο.

Censorinus, de D. N. 17.
Compare Plutarch, Plac. Philos. v. 24, p. 909.
LXXXVIII.—Crameri A. P. i. p. 324.
Compare Philo, Qu. on Gen. ii. 5, p. 82 Aucher.
Plutarch, de Orac. def. 12, p. 416.
LXXXIX.—Philo, Qu. in Gen. ii. 5, p. 82 Aucher.
XCI.—Stobaeus Floril. iii. 84.
Compare Cleanthes H., Iov. 24.
Hippocrates, π. τροφῆς 15. Plutarch, de Iside 45, p. 369.
Plotinus, Enn. vi. 5, p. 668. Empedocles v. 231 Stein.

a city holds fast to its law. For all human laws are dependent upon one divine Law, for this rules as far as it wills, and suffices for all, and overabounds.

XCII.—Although the Law of Reason is common, the majority of people live as though they had an understanding of their own.

XCIII.—They are at variance with that with which they are in most continual association.

XCIV.—We ought not to act and speak as though we were asleep.

XCV.—Plutarch, de Superst. 3, p. 166. Heraclitus says: To those who are awake, there is one world in common, but of those who are asleep, each is withdrawn to a private world of his own.

XCVI.—For human nature does not possess understanding, but the divine does.

XCII.—Sextus Emp. adv. Math. vii. 133. Context:—For having thus statedly shown that we do and think everything by participation in the divine reason, he (Heraclitus), after some previous exposition, adds: It is necessary, therefore, to follow the common (for by ξυνὸς he means ὁ κοινός, the common). For although the law of reason is common, the majority of people live as though they had an understanding of their own. But this is nothing else than an explanation of the mode of the universal disposition. As far, therefore, as we participate in the memory of this, we are true ; but in as far as we act individually, we are false.

XCIII.—M. Antoninus iv. 46. Context, see frag. 5.

XCIV.—M. Antoninus iv. 46. Context, see frag. 5.

XCV.—Compare pseudo-Pythagoras from Hippolytus, Ref. haer. vi. 26.

Iamblichus, Protrept. 21, p. 132, Arcer.

XCVI.—Origen, c. Cels. vi. 12, p. 291. Context:—Nevertheless he (Celsus) wanted to show that this was a fabrication of ours and taken from the Greek philosophers, who say that human wisdom is of one kind, and divine wisdom of another. And he brings forward some phrases of Heraclitus, one where he says, "For human nature does not possess understanding, but the divine does" And another, "The thoughtless man understands the voice of the Deity as little as the child understands the man" (= frag. 97).

XCVII.—The thoughtless man understands the voice of the Deity as little as the child understands the man.[32]

XCVIII.—Plato, Hipp. mai. 289 B. And does not Heraclitus, whom you bring forward, say the same, that the wisest of men compared with God appears an ape in wisdom and in beauty and in all other things?

XCIX.—Plato, Hipp. mai. 289 A. You are ignorant, my man, that there is a good saying of Heraclitus, to the effect that the most beautiful of apes is ugly when compared with another kind, and the most beautiful of earthen pots is ugly when compared with maidenkind, as says Hippias the wise.

C.—The people must fight for their law as for their walls.

CI.—Greater fates gain greater rewards.

CII.—Gods and men honor those slain in war.

CIII.—Presumption must be quenched even more than a fire.[33]

XCVII —Origen, c Cels. vi. 12, p. 291. Context, see frag 96.
Compare M Antoninus iv 46. Context, see frag 5.

XCVIII —Compare M Antoninus iv. 16.

XCIX —Compare Plotinus, Enn. vi. 3, p. 626.
Aristotle, Top. iii. 2, p. 117 b 17.

C —Diogenes Laert. ix. 2. Context.—And he (Heraclitus) used to say, "It is more necessary to quench insolence than a fire" (= frag. 103) And, "The people must fight for their law as for their walls."

CI.—Clement of Alex. Strom. iv. 7, p. 586. Context —Again Aeschylus, grasping this thought, says, "To him who toils, glory from the gods is due as product of his toil " " For greater fates gain greater rewards," according to Heraclitus.
Theodoretus, Therap. viii. p. 117, 33.
Compare Hippolytus, Ref. haer. v. 8.

CII.—Clement of Alex Strom iv. 4, p. 571. Context —Heraclitus said, "Gods and men honor those slain in war."
Theodoretus, Therap viii p. 117, 33.

CIII.—Diogenes Laert. ix. 2. Context, see frag. 100.

ON NATURE.

CIV.—For men to have whatever they wish, would not be well. Sickness makes health pleasant and good; hunger, satiety; weariness, rest.

CV.—It is hard to contend against passion, for whatever it craves it buys with its life.

CVI.—†It pertains to all men to know themselves and to learn self-control.†

CVII.—†Self-control is the highest virtue, and wisdom is to speak truth and consciously to act according to nature.†[34]

CVIII.—It is better to conceal ignorance, but it is hard to do so in relaxation and over wine.

CIV.—Stobaeus Floril. iii. 83, 4.
Compare Clement of Alex Strom. ii. 21, p. 497.
Theodoretus, Therap. xi. p. 152, 25. Context:—Heraclitus the Ephesian changed the name but retained the idea, for in the place of pleasure he put contentment.

CV.—Iamblichus, Protrept. p. 140, Arcer. Context:—Heraclitus is a witness to these statements, for he says, "It is hard to contend against passion," etc
Aristotle, Eth. Nic. ii. 2, p. 1105 a 8.
Idem, Eth. Eud. ii. 7, p. 1223 b 22.
Idem, Pol. v. 11, p. 1315 a 29.
Plutarch, de Cohib. ira 9, p. 457.
Idem, Erot. 11, p. 755.
Compare Plutarch, Coriol. 22.
Pseudo-Democritus fr. mor. 77, Mullach.
Longinus, de Subl. 44.

CVI.—Stobaeus Floril. v. 119.

CVII.—Stobaeus Floril. iii. 84.

CVIII.—Plutarch, Qu. Conviv. iii. proem., p. 644. Context:—Simonides, the poet, seeing a guest sitting silent at a feast and conversing with no one, said, "Sir, if you are foolish you are doing wisely, but if wise, foolishly," for, as Heraclitus says, "It is better to conceal ignorance, but it is hard," etc.
Idem, de Audiendo 12, p. 43.
Idem, Virt. doc. posse 2, p 439.
Idem, from Stob. Floril. xviii. 32.

CIX.—†It is better to conceal ignorance than to expose it.†

CX.—It is law, also, to obey the will of one.[35]

CXI.—For what sense or understanding have they? They follow minstrels and take the multitude for a teacher, not knowing that many are bad and few good. For the best men choose one thing above all—immortal glory among mortals; but the masses stuff themselves like cattle.

CXII.—In Priene there lived Bias, son of Teutamus, whose word was worth more than that of others.

CXIII.—To me, one is ten thousand if he be the best.

CXIV.—The Ephesians deserve, man for man, to be hung, and the youth to leave the city, inasmuch as they have banished Hermodorus, the worthiest man among them, saying: "Let no one of us excel, and if

---

CIX.—Stobaeus Floril. iii. 82.

CX.—Clement of Alex. Strom. v. 14, p. 718 (Euseb. P. E. xiii. 13, p. 681). Context, see frag. 65.

CXI.—The passage is restored as above by Bernays (Heraclitea i. p. 34), and Bywater (p. 43), from the following sources:

Clement of Alex. Strom. v. 9, p. 682.

Proclus in Alcib. p. 255 Creuzer, = 525 ed. Cous. ii.

Clement of Alex. Strom. iv. 7, p. 586.

CXII.—Diogenes Laert. i. 88. Context:—And the fault-finding Heraclitus has especially praised him (Bias), writing, "In Priene there lived Bias, son of Teutamus, whose word was worth more than that of others," and the Prienians dedicated to him a grove called the Teutamion. He used to say, "Most men are bad."

CXIII.—Theodorus Prodromus in Lazerii Miscell. i. p. 20.

Idem, Tetrastich. in Basil. I (fol. κ 2 vers. ed. Bas.).

Galenus, περὶ διαγνώσεως σφυγμῶν i. 1; t. 3, p. 53 ed. Bas.

Symmachus, Epist. ix. 115.

Compare Epigramm. from Diogenes Laert. ix. 16.

Cicero, ad. Att. xvi. 11.

Seneca, Epist. 7.

CXIV.—Strabo xiv. 25, p. 642. Context:—Among distinguished men of the ancients who lived here (Ephesus) were Heraclitus,

there be any such, let him go elsewhere and among other people."

CXV.—Dogs, also, bark at what they do not know.

CXVI.—By its incredibility, it escapes their knowledge.[36]

CXVII.—A stupid man loves to be puzzled by every discourse.

CXVIII.—The most approved of those who are of repute knows how to cheat. Nevertheless, justice will catch the makers and witnesses of lies.[37]

CXIX.—Diogenes Laert. ix. 1. And he (Heraclitus) called the obscure, and Hermodorus, of whom Heraclitus himself said, "The Ephesians deserve," etc.
Cicero, Tusc. v. 105.
Musonius from Stob. Floril. xl. 9.
Diogenes Laert. ix. 2.
Iamblichus, de Vit. Pyth. 30, p. 154 Arcer.
Compare Lucian, Vit. auct. 14.
Pseudo-Diogenes, Epist. 28, 6.

CXV.—Plutarch, An seni sit ger. resp. vii. p. 787. Context:—And envy, which is the greatest evil public men have to contend with, is least directed against old men. "For dogs, indeed, bark at what they do not know," according to Heraclitus.

CXVI.—Plutarch, Coriol. 38. Context:—But knowledge of divine things escapes them, for the most part, because of its incredibility, according to Heraclitus.
Clement of Alex. Strom. v. 13, p. 699. Context, see Crit. Note 36.

CXVII.—Plutarch, de Audiendo 7, p. 41. Context:—They reproach Heraclitus for saying, "A stupid man loves," etc.
Compare idem, de Aud. poet. 9, p. 28.

CXVIII.—Clement of Alex. Strom. v. 1, p. 649. Context:—"The most approved of those who are of repute knows how to be on his guard (φυλάσσειν, see Crit. Note 37). Nevertheless, justice will catch the makers and witnesses of lies," says the Ephesian. For this man who was acquainted with the barbarian philosophy, knew of the purification by fire of those who had lived evil lives, which afterwards the Stoics called the conflagration (ἐκπύρωσιν).

CXIX.—Schleiermacher compares Schol. Ven. ad Iliad xviii. 251 and Eustathius, p. 1142, 5 ed. Rom., which, however, Bywater does not regard as referring to Heraclitus of Ephesus.

used to say that Homer deserved to be driven out of the lists and flogged, and Archilochus likewise.

CXX.—One day is like all.

CXXI.—A man's character is his daemon.[38]

CXXII.—There awaits men after death what they neither hope nor think.

CXXIII.—And those that are there shall arise and become guardians of the living and the dead.[39]

CXXIV.—Night-roamers, Magians, bacchanals, revelers in wine, the initiated.

CXX.—Seneca, Epist. 12. Context —Heraclitus, who got a nickname for the obscurity of his writing, said, "One day is like all." His meaning is variously understood. If he meant all days were equal in number of hours, he spoke truly But others say one day is equal to all in character, for in the longest space of time you would find nothing that is not in one day, both light and night and alternate revolutions of the earth.

Plutarch, Camill. 19. Context —Concerning unlucky days, whether we should suppose there are such, and whether Heraclitus did right in reproaching Hesiod who distinguished good and bad days, as being ignorant that the nature of every day is one, has been examined in another place.

CXXI —Plutarch, Qu. Platon. i. 2, p. 999. Context.—Did he, therefore (viz. Socrates) call his own nature, which was very critical and productive, God? Just as Menander says, "Our mind is God." And Heraclitus, "A man's character is his dæmon."

Alexander Aphrod. de Fato 6, p. 16, Orell

Stobaeus Floril. civ 23. Comp pseudo-Heraclitus, Epist. 9.

CXXII —Clement of Alex Strom. iv. 22, p 630 Context:—With him (Socrates), Heraclitus seems to agree when he says in his discourse on men, "There awaits men," etc.

Idem, Protrept 2, p 18. Theodoretus, Therap. viii. p. 118, 1.

Themistius (Plutarch) from Stob. Floril. cxx 28.

CXXIII.—Hippolytus, Ref. haer. ix. 10. Context.—And he (Heraclitus) says also that there is a resurrection of this visible flesh of ours, and he knows that God is the cause of this resurrection, since he says, "And those that are there shall arise," etc.

Compare Clement of Alex. Strom v. 1, p. 649.

CXXIV —Clement of Alex. Protrept. 2, p. 18. Context:—Rites worthy of the night and of fire, and of the great-hearted, or rather

CXXV.—For the things which are considered mysteries among men, they celebrate sacrilegiously.

CXXVI.—And to these images they pray, as if one should prattle with the houses knowing nothing of gods or heroes, who they are.

CXXVII.—For were it not Dionysus to whom they institute a procession and sing songs in honor of the pudenda, it would be the most shameful action. But Dionysus, in whose honor they rave in bacchic frenzy, and Hades are the same.[40]

CXXVIII.—Iamblichus, de Mysteriis v. 15. I distinguish two kinds of sacrifices. First, those of men wholly purified, such as would rarely happen in the case of a single individual, as Heraclitus says, or of a

---

of the idle-minded people of the Erechthidae, or even of the other Greeks, for whom there awaits after death what they do not hope (see frag. 122). Against whom, indeed, does Heraclitus of Ephesus prophesy? Against night-roamers, Magians, bacchanals, revelers in wine, the initiated. These he threatens with things after death and prophesies fire for them, for they celebrate sacrilegiously the things which are considered mysteries among men (= frag. 125).

CXXV.—Clement of Alex. Protrept. 2, p. 19. Context, see frag. 124.

Compare Arnobius, adv. Nat. v 29.

CXXVI.—Origen, c. Cels. vii. 62, p. 384.

Idem i. 5, p. 6.

Clement of Alex. Protrept. 4, p. 44. Context:—But if you will not listen to the prophetess, hear your own philosopher, Heraclitus, the Ephesian, imputing unconsciousness to images, "And to these images," etc.

CXXVII.—Clement of Alex. Protrept. 2, p. 30. Context:—In mystic celebration of this incident, phalloi are carried through the cities in honor of Dionysus. "For were it not Dionysus to whom they institute a procession and sing songs in honor of the pudenda, it would be the most shamful action," says Heraclitus. "But Hades and Dionysus are the same, to whom they rave in bacchic frenzy," not for the intoxication of the body, as I think, so much as for the shameful ceremonial of lasciviousness.

Plutarch, de Iside 28, p. 362.

certain very few men. Second, material and corporeal sacrifices and those arising from change, such as are fit for those still fettered by the body.

CXXIX.—Atonements.[41]

CXXX.—When defiled, they purify themselves with blood, just as if any one who had fallen into the mud should wash himself with mud!

CXXIX —Iamblichus, de Mys i. 11. Context·—Therefore Heraclitus rightly called them (scil. what are offered to the gods) "atonements," since they are to make amends for evils and render the souls free from the dangers in generation

Compare Hom Od xxii 481. See Crit. Note 41

CXXX —Elias Cretensis in Greg. Naz. l. l. (cod. Vat. Pii. 11, 6, fol. 90 r). Context:—And Heraclitus, making sport of these people, says, "When defiled, they purify themselves with blood, just as if any one who had fallen into the mud should wash himself with mud!" For to suppose that with the bodies and blood of the unreasoning animals which they offer to their gods they can cleanse the impurities of their own bodies, which are stained with vile contaminations, is like trying to wash off mud from their bodies by means of mud

Gregory Naz. Or. xxv. (xxiii) 15, p. 466 ed. Par. 1778.

Apollonius, Epist. 27.

Compare Plotinus, Enn. i. 6, p. 54.

CRITICAL NOTES.

FRAGMENT 1.

Note 1.—Instead of λόγου, MS has δόγματος, corrected by Bernays, followed by all critics except Bergk.

FRAGMENT 2.

Note 2.—The λόγος of Heraclitus stood for the element of order or law in the ever-shifting world. Our word Reason may express the same idea more in accord with the thought of that time (see Introduction, p. 59 ff.). Zeller and Pfleiderer understand by it, Reason ruling or immanent in the world; Heinze, the objective (unconscious) law of Reason; Bernays, conscious Intelligence; Teichmüller, self-conscious Reason; Schuster, on the other hand, regards it as the "revelation offered us by the audible Speech of Nature." In the present passage, Zeller is inclined to understand by τοῦ λόγου τοῦδε, primarily the discourse of the author, but containing also the idea of the content of the discourse, i. e. the theory of the world laid down in his book (Vol 1, p. 572, 2). For fuller account of the λόγος, compare Introduction, pp. 8, 12, 28, 45, 59, 61.

FRAGMENT 13.

Note 3.—Bywater reads, Ὅσων ὄψις ἀκοὴ μάθησις, ταῦτα ἐγὼ προτιμέω; Compare Introduction, p. 19 f.

FRAGMENT 15.

Note 4.—Compare Introduction, p. 48. Bernays (Rhein. Mus. ix. 261 f.) offers the explanation that the eyes are more exact witnesses than the ears, because by the eyes we have the only pure cognition of fire, in the perception of which is the only true knowledge.

FRAGMENT 18.

Note 5.—See Introduction, p. 36 ff.

FRAGMENT 19.

Note 6.—Common reading has ἓν τὸ σοφόν ἐπίστασθαι γνώμην ἧτε οἱ ἐγκυβερνήσει πάντα διὰ πάντων. Schleiermacher, γνώμην οἴη κυβερνήσει. Bernays, ἧτε οἰακίζει. Schuster, ἧτε οἴη τε κυβερνήσει.

FRAGMENT 20.

Note 7.—The sense of ἁπάντων is uncertain. In the citations from Plutarch and Simplicius, the word is omitted; they read κόσμον τονδε. Zeller, whose interpretation of the word we have followed, takes it as masculine, referring to the gods and men, the meaning then being, that since gods and men are included in the world as part of it, they could not have created it. Schuster, on the other hand, renders it as follows: "Die Welt, die alles in sich befasst [die neben sich weder für andre Welten noch für einen Schöpfer Raum hat]," etc.

FRAGMENT 21.

Note 8.—Πρηστήρ is rendered by Schuster "fiery wind" such as forms the stars. Zeller (Vol. 1, p. 588, 1) believes it has essentially the same signification as κεραυνός in frag. 28, both words being other terms for the world-ruling fire or formative principle of the world.

FRAGMENT 23.

Note 9.—Eusebius omits γῆ, and is followed by Lassalle and Heinze. The former (Vol. 2, p. 63) translates, "Das Meer wird ausgegossen und gemessen nach demselben Logos, welcher zuerst war, ehe es (selbst) noch war," and finds here a confirmation of his interpretation of the Logos as the eternal preëxisting law of the identity of being and not-being. Heinze understands it as follows: "Das Meer verwandelt sich in denselben Logos, also in dasselbe Feuer, von welcher Beschaffenheit es vorher war, ehe es selbst entstand." Schuster reads γῆν and translates, "Das Meer ergiesst sich und nimmt sein Maass ein im selben Umfang, wie damals als noch keine Erde war" (p. 129). Zeller reads γῆ and understands the passage to refer to the return of the earth into the sea from which it sprang. By λόγος here he understands "proportion of magnitude" or "size," so that ἐς τὸν αὐτὸν λόγον means that the sea returns "to the same size" as before it became earth (Vol. 1, p. 628, 3).

FRAGMENT 24.

Note 10.—See Introduction, pp. 15, 22, 68.

FRAGMENT 25.

Note 11.—This fragment is not accepted by Zeller, who holds that air was not recognized by Heraclitus as one of the elements, but that he accepted only the three, fire, water, and earth. Air was added, Zeller thinks, by later writers, who confused it with the "soul" of Heraclitus (Vol. 1, p. 615). Schuster, who thinks Heraclitus did not teach a specific number of elements after the manner of Empedocles, regards the passage as trustworthy (p. 157 ff.). Teichmüller gives to air an important place in the system of Heraclitus, distinguishing the upper pure air, which is not different from fire, and the impure lower air (Vol. 1, p. 62).

FRAGMENT 27.

Note 12.—Schleiermacher, followed by Mullach, reads τινα for τις, so that the sense becomes, "How can that which never sets escape any one?" This is unnecessary and violates the context in Clement. That which never sets is the eternal Order or Law, conceived here as Destiny or Justice According to Zeller (Vol. 1, p. 590), that which never sets is fire. According to Schuster (p. 184), it is Relation or Law, and the τις refers to Helios, which, though itself the centre of power and intelligence, is yet subject to law. Teichmüller (Vol. 1, p. 184) understands it to refer to Justice or Destiny, which never sets like the sun, and which none can escape.

FRAGMENT 35

Note 13.—Πλείστων may be taken as neuter: "Hesiod was a teacher of the greatest number of things." On the unity of day and night, compare Introduction, p. 32 f.

FRAGMENT 36.

Note 14.—The original text, which reads ὁκόταν συμμιγῇ θυώμασι, has been variously corrected. As the subject of συμμιγῇ, Schuster inserts οἶνος, the sense then being that as wine is mixed with spices and labelled as any one pleases, so God receives different names under different forms (p. 188). Bywater, following Bernays (Rhein. Mus. ix. 245), inserts θίωμα, and Zeller (Vol. 1, p 602, 2) reads ὅκως ἀήρ for ὁκωσπερ. Teichmuller (Vol. 1, p. 67) attempts to save the original reading by making ὁ θεός, (i. e. fire) the subject both of ἀλλοιοῦται and συμμιγῇ. The correction of Bernays is the most satisfactory; the meaning then being, that as when perfumes are mixed, the mixture is named according to the scent that impresses each person, so God is named according to the attribute that most impresses the individual. Compare frag. 65. About the same sense, however, is derived from the other readings.

FRAGMENT 38.

Note 15.—Schleiermacher and Zeller think it doubtful whether any sense can be made out of this fragment. For Schuster's fanciful explanation, see Introduction, p. 18 f. Bernays (Rhein. Mus. ix. p. 265, 6) interprets it to mean that the perception of fire, upon which depends the existence of the soul, is gained after death and the extinction of the sense of sight, by the sense of smell, just as the passage from Aristotle (frag. 37) teaches that in the conflagration of the world, all perception will be by the nostrils. Pfleiderer (p. 218) suggests ὀσιοῦνται for ὀσμῶνται.

FRAGMENT 40.

Note 16.—Of this passage from Plutarch only the words σκίδνησι καὶ συνάγει, πρόσεισι καὶ ἄπεισι, can with any certainty be attributed directly to Heraclitus. The rest bears marks of later hands, as shown by Bernays (Heraklit. Briefe, p. 55), and Zeller (Vol. 1, p. 576, 2).

FRAGMENT 45.

Note 17.—Bernays' explanation of this passage (Rhein Mus. vii. p 94 ; compare Introduction, p. 44 f) has been followed by Zeller, Schuster (partly), and Arnold Hug. According to this interpretation, the association of the bow and lyre lies in their form, which in the case of the old Greek or Scythian bow with its arms bent back at the ends, was like that of the lyre. Hence we have in the bow and the lyre, two distinct illustrations of harmony by opposite straining tension. Lassalle (Vol. 1, p. 113) understands it to refer to the harmony *between* the bow and the lyre; the bow and the lyre being symbols in the Apollo cult, the one of singularity and difference, the other, of universality and union. On Pfleiderer's modification of Lassalle's view, see Introduction, p. 44. In place of τόξου καὶ λύρης, Bast reads τοῦ ὀξέος τε καὶ βαρέος. Bergk conjectures τόξου καὶ νευρῆς. On the interpretation of this passage by Plutarch and Plato's Eryximachus as the harmony of sharps and flats in music, compare Hug (Platons Symposion, p. 77, 5) and Zeller (Vol. 1, p 578, 2). Compare frags. 56, 43, 59.

FRAGMENT 47.

Note 18.—Schuster (p 24, note) reads ἐς τί γὰρ φησίν, ἁρμονίη ἀφανὴς φανερῆς κρείττων. See Introduction, p. 20, and Zeller, Vol. 1, p. 604, 1.

FRAGMENT 50.

Note 19.—MS reads γραφέων ; Duncker and Bywater, γναφέων ; Bernays, γναφείῳ.

FRAGMENT 55.

Note 20 —The common reading is πᾶν ἑρπετὸν τὴν γῆν νέμεται, which Zeller retains, understanding it to refer to the beastliness of men, who "feed upon the earth like the worm" (Vol. 1, p 660). Pfleiderer likewise accepts this reading, quoting Sallust, Catil. 1 : Vitam silentio transeunt veluti pecora, quae natura prona atque ventri obedientia finxit. That πληγῇ, the reading of Stobaeus, followed by Bywater, is correct, however, is shown by comparison with Æschylus, Ag. 358, Διὸς πλαγὰν ἔχουσιν εἰπεῖν, and Plato's Criti. 109 B, καθάπερ ποιμένες κτήνη πληγῇ νέμοντες. With this reading, the sense then becomes that man is subject to eternal divine force or law.

Fragment 56.

Note 21.—Compare frag. 45 and note 17. Bywater reads παλίντονος ἁρμονίη, here; but though in three passages, those namely given under this fragment, παλίντονος is found in the MSS, yet the context even in Plutarch, where sharps and flats are spoken of, calls for the meaning "harmony of oppositions," as explained in note 17, for which we should expect παλίντροπος rather than παλίντονος.

Fragment 60.

Note 22.—What is referred to by ταῦτα, "these things," has been questioned. Teichmüller, followed by Pfleiderer, has given the true explanation. Ταῦτα refers to some idea the opposite of "justice." Clement is illustrating the Pauline principle that without law there would have been no sin. For this, Heraclitus, whose prominent thought was, no war without peace, no good without bad, etc., served him as good authority.

Fragment 62.

Note 23.—The original text is as follows: Εἰ δὲ χρὴ τὸν πόλεμον ἐόντα ξυνὸν καὶ δίκην ἐρεῖν καὶ γινόμενα πάντα κατ' ἔριν καὶ χρεώμενα. Schleiermacher proposes εἰδέναι for εἰ δέ and ἔριν for ἐρεῖν, and has been followed by Zeller, Bywater and others. Schuster retains the MS form in the first clause. Χρεώμενα also gives trouble Brandis proposes σωζόμενα. Schuster reads καταχρεώμενα, approved by Zeller. Lassalle and Bywater retain χρεώμενα. This passive use is unusual, but possible, as shown by the analogy of καταχρεώμενα. The translations of Schuster and Lassalle are as follows.

Schuster (p. 198)—"In dem Falle muss man also den gemeinsamen Krieg sogar Recht nennen und [sagen] das alles [nur] in Folge des Streites entsteht und sich aufbraucht."

Lassalle—"Man muss wissen dass der Krieg das Gemeinsam ist, und der Streit das Recht, und dass nach dem Gesetz des Streits alles wird und verwendet wird (or lit. und sich bethätigt)."

Ξυνός in this passage has almost the signification "common good."

Fragment 64.

Note 24.—Critics have expended their ingenuity in trying to make something out of this obscure fragment. Teichmuller (Vol. 1, p 97 ff.) says that we have here the distinction of the intelligible from the sensible world. The former is the pure, light, fiery and most incorporeal being, compared with which the world of the senses is death. Zeller (Vol. 1, p. 651) similarly refers it to the testimony of the senses, which see the world as something "stiff and dead," when really everything is in constant motion. Schuster (p. 276) labors with a far-fetched interpretation to show that the passage does *not* cast any disparagement upon the senses. For Pfleiderer's explanation, see Introduction, p. 43. All these interpretations look for a theoretical meaning, when it is quite possible that no theoretical meaning was intended. It is simpler to compare it with frag. 2, and refer it to Heraclitus' repeated charge against the people, of their sleep-like condition when awake.

Fragment 65.

Note 25.—We have followed Schuster's punctuation of this fragment Bywater, with other critics, reads, Ἐν τὸ σοφὸν μοῦνον λέγεσθαι οὐκ ἐθέλει καὶ ἐθέλει Ζηνὸς οὔνομα. Τὸ σοφόν, here, is the world-ruling Wisdom or Order, to which Heraclitus applies many names. (See Introduction, p. 60 f.) It wills and wills not to be called by the name of Zeus, because that name, while it points towards its true nature, yet but partly indicates it, or in part wrongly. The variety of meanings, however, which have been drawn from this fragment may be shown by the following translations. Schleiermacher (and Lassalle). "Das Eine Weise allein will nicht ausgesprochen werden und will ausgesprochen werden, der Name des Zeus." Schuster: "Nur eines ist die Weisheit; sie lasst sich nicht und lasst sich doch auch wieder benennen mit des Zeus Namen." Bernays· "Eines, das allein Weise, will und will auch nicht mit des Ζῆν Namen genannt werden." The poetical form Ζηνὸς is chosen, thinks Bernays, to indicate that the One Wise is the source of "life." Zeller: "Eines, das allein Weise, will und will auch nicht mit dem Namen des Zeus benannt werden" Pfleiderer "Als Eins will das weise Allwesen, Zeus genannt, nicht bezeichnet werden und will es." Teichmuller: "Die Weisheit, Zeus genannt, will allein eins heissen und will es auch nicht."

Fragment 72.

Note 26.—This fragment is connected by Schuster and Zeller with the group of passages concerning rest in change (see frags. 82, 83), and refers to the pleasure which the rest and change of death bring to souls They therefore reject the μὴ θάνατον of Numenius as not Heraclitic. (Schuster, p. 191, 1 Zeller, p 647, 2) Pfleiderer, however (p. 222), retains the μὴ θάνατον as genuine, and explains that it is a pleasure to souls to become wet, because so by pursuing the way down into apparent death, they attain their new birth of life in death. He therefore retains also the τέρψιν δὲ εἶναι αὐταῖς τὴν εἰς τὴν γένεσιν πτῶσιν, of Numenius, as expressing the true sense of the passage.

Fragment 74.

Note 27 —The added clause of Plutarch, "It flashes through the body like lightning through the clouds," is also regarded by Schleiermacher, Schuster, Zeller, and Pfleiderer, as Heraclitic.

The similarity of the three fragments 74, 75, and 76 suggests, of course, that they are all corrupted forms of a common original. Bywater, however, accepts the form of expression in frag. 74 as surely Heraclitic and marks the other two as doubtful. Schleiermacher, from the number of citations of each of these fragments, concludes that Heraclitus had expressed himself in each of these three forms. Lassalle, in agreeing with him, believes also that Heraclitus, who was given to playing upon words (for further examples of Heraclitus' puns, compare frags. 91, 101, 127, 66), not without purpose chose the words αὔη and αὐγή, and sees in the use of the latter word a reference to the lightning-like movement of the soul (Vol. 2, p. 196 f.). Zeller thinks it difficult to determine the original form, but he does not regard the proposition αὐγὴ ξηρὴ ψυχὴ σοφωτάτη, as Heraclitic (Vol. 1, p. 643, 2).

Fragment 77.

Note 28.—The original of this difficult and corrupted passage as it appears in Clement, is as follows (unpunctuated), Ἄνθρωπος ἐν εὐφρόνῃ φάος ἅπτεται ἑαυτῷ ἀποθανὼν ἀποσβεσθεὶς ζῶν δὲ ἅπτεται τεθνεῶτος εὕδων ἀποσβεσθεὶς ὄψεις ἐγρηγορὼς ἅπτεται εὕδοντος. Various emendations and translations of this have been made. Compare Schuster, p. 271; Pfleiderer, p. 204, 1. Bywater, however, finally rescues as Heraclitic the form given above in the text.

Fragment 80.

Note 29.—That this fragment is to be taken in the sense in which Diogenes understands it, rather than in that of Plutarch, is held by Schuster (p. 61) and Zeller (Vol. 1, p. 654, 4). Lassalle (Vol. 1, p. 301), following Schleiermacher, takes it as Clement does, in the sense of the Delphic inscription, "I have sought myself in the general flux of things, I have striven to know myself." For Pfleiderer's interpretation and the true meaning, see Introduction, pp. 41, 48.

Fragment 82.

Note 30.—Lassalle, following Creuzer, reads ἄγχεσθαι instead of ἄρχεσθαι (Vol. 1, p. 131.)

Fragment 90.

Note 31.—Lassalle (Vol. 1, p. 290) interprets this fragment as follows: In waking, we distinguish our own representations from the objective world common to all. In sleeping, they are one and the same. Hence Heraclitus says the sleeping make their own world. Similarly Pfleiderer (p. 202 f.) understands Heraclitus to mean that the sleeper makes his own world, while the waking man is conscious that corresponding to his world of ideas there is a common objective world. Pfleiderer rejects καὶ συνεργοὺς as an addition of Aurelius.

FRAGMENT 97.

Note 32.—This fragment has given trouble. Bernays (Heraclitea 15) proposes to substitute δαήμονος for δαίμονος, but has not been followed by other critics. Schleiermacher translates, "Ein thörichter Mann vernimmt nicht mehr von Schicksal als ein Kind von einem Mann." Schuster (p. 342) renders, "Der Mensch in seiner Kindheit hat (sie [*i. e.* the names]) von Gott gehört, wie (jetzt) das Kind von dem Manne," and finds here support for the theory of the natural fitness of names (see Introduction, p. 16), which primitive man learned directly from Nature. Zeller (Vol. 1, p. 653) refers it to the childish want of reason in man, which does not perceive the voice of the deity. Pfleiderer (p. 51) renders, "Der unverständige Mensch hat von jeher nur soviel von der Gottheit gehört, als ein Kind vom Manne."

FRAGMENT 103.

Note 33.—Ὕβριν here is to be taken in the sense of excess of self-assertion, the private will against the universal Law. Compare frags. 92, 104, etc.

FRAGMENT 107.

Note 34.—The latter clause may also be translated, "Wisdom is to speak and act truly, giving ear to Nature."

FRAGMENT 110.

Note 35.—Clementine MS reads βουλή. Eusebius, followed by all but Mullach, reads βουλῇ. For Heraclitus' opinions on democracy, see, further, frags. 114, 113.

FRAGMENT 116.

Note 36.—The passage in Clement is as follows: ἀλλὰ τὰ μὲν τῆς γνώσεως βάθη κρύπτειν ἀπιστίη ἀγαθή, καθ' Ἡράκλειτον· ἀπιστίη γὰρ διαφυγγάνει μὴ γιγνώσκεσθαι, from which it is seen that the words of Heraclitus, ἀπιστίη διαφυγγάνει μὴ γιγνώσκεσθαι, were differently understood by Clement and Plutarch. Schuster (p. 72) accepts the Clementine form, and regards the whole passage as Heraclitic, and renders, "Die Tiefe der Erkenntniss zu verbergen, das ist ein gutes Misstrauen. Denn durch diese misstrauische Behutsamkeit entgeht man dem Schicksal durchschaut zu werden," by which he accounts for the (intentional) obscurity of Heraclitus' writings. Zeller (Vol. 1, p. 574, 2), following Schleiermacher, rejects the Clementine version, and regards the words as teaching that truth is hidden from the masses because it seems incredible to them. A still different meaning may be found in the words if we take ἀπιστίη as subjective, referring to the want of faith which prevents us from seeing truth.

FRAGMENT 118.

Note 37.—The common reading is, δοκεόντων ὁ δοκιμώτατος γινώσκει φυλάσσειν, which makes nonsense. Schleiermacher proposes δοκέοντα ὁ δοκιμώτατος γινώσκειν φυλάσσειν. Schuster (p. 340) suggests, δοκεόντων, ὁ δοκιμώτατον γίεται, γινώσκει φυλάσσειν, and fancies the allusion is to the poets, who from credible things accept that which is most credible. Bergk, followed by Pfleiderer, reads φλνάσσειν, to talk nonsense. Bernays, followed by Bywater, reads πλάσσειν.

FRAGMENT 121.

Note 38.—This fragment has been variously translated, but the meaning seems to be that a man's God or Destiny depends not upon external divine powers, but upon his own inner nature. Teichmuller finds here the further meaning that the essence of mind is the essence of deity.

FRAGMENT 123.

Note 39.—The meaning of this passage is very doubtful. We have followed Bernays' reading instead of the common ἔνθα δεόντι, which Bywater retains, although he marks it uncertain. Schuster (p. 176, 1) suggests [δαίμων ἐθέλει] ἐνθάδε ἐόντι ἐπιίστασθαι καὶ φυλακός κ. τ. λ. Zeller (Vol. 1, p. 648, 4) regards it as a reference to the dæmons who are made protectors of men. Lassalle (Vol. 1, p. 185) thinks it refers to a resurrection of souls

FRAGMENT 127.

Note 40.—For text and discussion of this passage, see Introduction, p. 52 ff. Teichmüller's interpretation of it is as follows: "Wenn es nicht Dionysus wäre, dem sie die Procession fuhren und dabei das Lied auf die Schamglieder singen, so wäre das Schamloseste ausgefuhrt Nun aber, ist Hades (der Sohn der Scham) derselbe wie Dionysus, dem sie rasen und Feste feiern." This means, says Teichmuller, that the shameful and the becoming are the same (Identification of opposites). For what is improper for men is proper for Dionysus, because he is the same as Hades, and Hades is the same as shame, which latter he attempts to prove from Plutarch, de Is. 29 b. Again, Dionysus and Hades are the same, because the former stands for the sun and the latter for the lower world, and as the sun is absorbed into the earth at night and generated therefrom in the morning, they must be essentially the same. (Neue Studien, Vol. 1, p. 25.)

FRAGMENT 129.

Note 41.—That the use of this term was ironical, is made probable by the following fragment.

ΗΡΑΚΛΕΙΤΟΥ ΕΦΕΣΙΟΥ

ΠΕΡΙ ΦΥΣΕΩΣ.

I. Οὐκ ἐμεῦ ἀλλὰ τοῦ λόγου ἀκούσαντας ὁμολογέειν σοφόν ἐστι, ἓν πάντα εἶναι.

II. Τοῦ δὲ λόγου τοῦδ' ἐόντος αἰεὶ ἀξύνετοι γίγνονται ἄνθρωποι καὶ πρόσθεν ἢ ἀκοῦσαι καὶ ἀκούσαντες τὸ πρῶτον. γινομένων γὰρ πάντων κατὰ τὸν λόγον τόνδε ἀπείροισι ἐοίκασι πειρώμενοι καὶ ἐπέων καὶ ἔργων τοιουτέων ὁκοίων ἐγὼ διηγεῦμαι, διαιρέων ἕκαστον κατὰ φύσιν καὶ φράζων ὅκως ἔχει. τοὺς δὲ ἄλλους ἀνθρώπους λανθάνει ὁκόσα ἐγερθέντες ποιέουσι, ὅκωσπερ ὁκόσα εὕδοντες ἐπιλανθάνονται.

III. Ἀξύνετοι ἀκούσαντες κωφοῖσι ἐοίκασι· φάτις αὐτοῖσι μαρτυρέει παρεόντας ἀπεῖναι.

IV. Κακοὶ μάρτυρες ἀνθρώποισι ὀφθαλμοὶ καὶ ὦτα, βαρβάρους ψυχὰς ἐχόντων.

V. Οὐ φρονέουσι τοιαῦτα πολλοὶ ὁκόσοισι ἐγκυρέουσι οὐδὲ μαθόντες γινώσκουσι, ἑωυτοῖσι δὲ δοκέουσι.

VI. Ἀκοῦσαι οὐκ ἐπιστάμενοι οὐδ' εἰπεῖν.

VII. Ἐὰν μὴ ἔλπηαι, ἀνέλπιστον οὐκ ἐξευρήσει, ἀνεξερεύνητον ἐὸν καὶ ἄπορον.

VIII. Χρυσὸν οἱ διζήμενοι γῆν πολλὴν ὀρύσσουσι καὶ εὑρίσκουσι ὀλίγον.

IX. Ἀγχιβασίην.

X. Φύσις κρύπτεσθαι φιλεῖ.

XI. Ὁ ἄναξ οὗ τὸ μαντεῖόν ἐστι τὸ ἐν Δελφοῖς, οὔτε λέγει οὔτε κρύπτει, ἀλλὰ σημαίνει.

XII. Σίβυλλα δὲ μαινομένῳ στόματι ἀγέλαστα καὶ ἀκαλλώπιστα καὶ ἀμύριστα φθεγγομένη χιλίων ἐτέων ἐξικνέεται τῇ φωνῇ διὰ τὸν θεόν.

XIII. Ὅσων ὄψις ἀκοὴ μάθησις, ταῦτα ἐγὼ προτιμέω.

XIV. Polybius iv. 40: τοῦτο γὰρ ἴδιόν ἐστι τῶν νῦν καιρῶν, ἐν οἷς πάντων πλωτῶν καὶ πορευτῶν γεγονότων οὐκ ἂν ἔτι πρέπον εἴη ποιηταῖς καὶ μυθογράφοις χρῆσθαι μάρτυσι περὶ τῶν αγνοουμένων, ὅπερ οἱ πρὸ ἡμῶν περὶ τῶν πλείστων, ἀπίστους ἀμφισβητουμένων παρεχόμενοι βεβαιωτὰς κατὰ τὸν Ἡράκλειτον.

XV. Ὀφθαλμοὶ τῶν ὤτων ἀκριβέστεροι μάρτυρες.

XVI. Πολυμαθίη νόον ἔχειν οὐ διδάσκει· Ἡσίοδον γὰρ ἂν ἐδίδαξε καὶ Πυθαγόρην αὐτίς τε Ξενοφάνεα καὶ Ἑκαταῖον.

XVII. Πυθαγόρης Μνησάρχου ἱστορίην ἤσκησε ἀνθρώπων μάλιστα πάντων. καὶ ἐκλεξάμενος ταύτας τὰς συγγραφὰς ἐποίησε ἑωυτοῦ σοφίην, πολυμαθίην, κακοτεχνίην.

XVIII. Ὁκόσων λόγους ἤκουσα οὐδεὶς ἀφικνέεται ἐς τοῦτο, ὥστε γινώσκειν ὅτι σοφόν ἐστι πάντων κεχωρισμένον.

XIX. Ἓν τὸ σοφόν, ἐπίστασθαι γνώμην ᾗ κυβερνᾶται πάντα διὰ πάντων.

XX. Κόσμον < τόνδε > τὸν αὐτὸν ἁπάντων οὔτε τις θεῶν οὔτε ἀνθρώπων ἐποίησε, ἀλλ' ἦν αἰεὶ καὶ ἔστι καὶ ἔσται πῦρ ἀείζωον, ἁπτόμενον μέτρα καὶ ἀποσβεννύμενον μέτρα.

XXI. Πυρὸς τροπαὶ πρῶτον θάλασσα· θαλάσσης δὲ τὸ μὲν ἥμισυ γῆ, τὸ δὲ ἥμισυ πρηστήρ.

XXII. Πυρὸς ἀνταμείβεται πάντα καὶ πῦρ ἁπάντων, ὥσπερ χρυσοῦ χρήματα καὶ χρημάτων χρυσός.

XXIII. Θάλασσα διαχέεται καὶ μετρέεται ἐς τὸν αὐτὸν λόγον ὁκοῖος πρόσθεν ἦν ἢ γενέσθαι †γῆ†.

XXIV. Χρησμοσύνη ... κόρος.

XXV. Ζῇ πῦρ τὸν γῆς θάνατον, καὶ ἀὴρ ζῇ τὸν πυρὸς θάνατον· ὕδωρ ζῇ τὸν ἀέρος θάνατον, γῆ τὸν ὕδατος.

XXVI. Πάντα τὸ πῦρ ἐπελθὸν κρινέει καὶ καταλήψεται.

XXVII. Τὸ μὴ δῦνόν ποτε πῶς ἄν τις λάθοι;

XXVIII. Τὰ δὲ πάντα οἰακίζει κεραυνός.

XXIX. Ἥλιος οὐχ ὑπερβήσεται μέτρα· εἰ δὲ μή, Ἐρινύες μιν δίκης ἐπίκουροι ἐξευρήσουσι.

XXX. Ἠοῦς καὶ ἑσπέρης τέρματα ἡ ἄρκτος, καὶ ἀντίον τῆς ἄρκτου οὖρος αἰθρίου Διός.

XXXI. Εἰ μὴ ἥλιος ἦν, εὐφρόνη ἂν ἦν.

XXXII. Νέος ἐφ' ἡμέρῃ ἥλιος.

XXXIII. Diogenes Laert. i. 23: δοκεῖ δὲ (scil. Θαλῆς) κατά τινας πρῶτος ἀστρολογῆσαι καὶ ἡλιακὰς ἐκλείψεις καὶ τροπὰς προειπεῖν, ὥς φησιν Εὔδημος ἐν τῇ περὶ τῶν ἀστρολογουμένων ἱστορίᾳ· ὅθεν αὐτὸν καὶ Ξενοφάνης καὶ Ἡρόδοτος θαυμάζει. μαρτυρεῖ δ' αὐτῷ καὶ Ἡράκλειτος καὶ Δημόκριτος.

XXXIV. Plutarchus Qu. Plat. viii. 4, p. 1007: οὕτως οὖν ἀναγκαίαν πρὸς τὸν οὐρανὸν ἔχων συμπλοκὴν καὶ συναρμογὴν ὁ χρόνος οὐχ ἁπλῶς ἐστι κίνησις ἀλλ', ὥσπερ εἴρηται, κίνησις ἐν τάξει μετρον ἐχούσῃ καὶ πέρατα καὶ περιόδους. ὧν ὁ ἥλιος ἐπιστάτης ὢν καὶ σκοπός, ὁρίζειν καὶ βραβεύειν καὶ ἀναδεικνύναι καὶ ἀναφαίνειν μεταβολὰς καὶ ὥρας αἳ πάντα φέρουσι, καθ' Ἡράκλειτον, οὐδὲ φαύλων οὐδὲ μικρῶν, ἀλλὰ τῶν μεγίστων καὶ κυριωτάτων τῷ ἡγεμόνι καὶ πρώτῳ θεῷ γίνεται συνεργός.

XXXV. Διδάσκαλος δὲ πλείστων Ἡσίοδος· τοῦτον ἐπίστανται πλεῖστα εἰδέναι, ὅστις ἡμέρην καὶ εὐφρόνην οὐκ ἐγίνωσκε· ἔστι γὰρ ἕν.

XXXVI. Ὁ θεὸς ἡμέρη εὐφρόνη, χειμὼν θέρος, πόλεμος εἰρήνη, κόρος λιμός· ἀλλοιοῦται δὲ ὅκωσπερ ὁκόταν συμμιγῇ < θυωμα > θυώμασι· ὀνομάζεται καθ' ἡδονὴν ἑκάστου.

XXXVII. Aristoteles de Sensu 5, p. 443 a 21: δοκεῖ δ' ἐνίοις ἡ καπνώδης ἀναθυμίασις εἶναι ὀσμή, οὖσα κοινὴ γῆς τε καὶ ἀέρος. καὶ πάντες ἐπιφέρονται ἐπὶ τοῦτο περὶ ὀσμῆς· διὸ καὶ Ἡράκλειτος οὕτως εἴρηκεν, ὡς εἰ πάντα τὰ ὄντα καπνὸς γένοιτο, ῥῖνες ἂν διαγνοῖεν.

XXXVIII. Αἱ ψυχαὶ ὀσμῶνται καθ' ᾅδην.

XXXIX. Τὰ ψυχρὰ θέρεται, θερμὸν ψύχεται, ὑγρὸν αὐαίνεται, καρφαλέον νοτίζεται.

XL. Σκίδνησι καὶ συνάγει, πρόσεισι καὶ ἄπεισι.

XLI. Ποταμοῖσι δὶς τοῖσι αὐτοῖσι οὐκ ἂν ἐμβαίης· ἕτερα γὰρ < καὶ ἕτερα > ἐπιρρέει ὕδατα.

XLII. † Ποταμοῖσι τοῖσι αὐτοῖσι ἐμβαίνουσιν ἕτερα καὶ ἕτερα ὕδατα ἐπιρρεῖ †.

XLIII. Aristoteles Eth. Eud. vii. ⊦, p. 1235 a 26: καὶ Ἡράκλειτος ἐπιτιμᾷ τῷ ποιήσαντι· ὡς ἔρις ἔκ τε θεῶν καὶ ἀνθρώπων ἀπόλοιτο· οὐ γὰρ ἂν εἶναι ἁρμονίαν μὴ ὄντος ὀξέος καὶ βαρέος, οὐδὲ τὰ ζῷα ἄνευ θήλεος καὶ ἄρρενος, ἐναντίων ὄντων.

XLIV. Πόλεμος πάντων μὲν πατήρ ἐστι πάντων δὲ βασιλεύς, καὶ τοὺς μὲν θεοὺς ἔδειξε τοὺς δὲ ἀνθρώπους, τοὺς μὲν δούλους ἐποίησε τοὺς δὲ ἐλευθέρους.

XLV. Οὐ ξυνίασι ὅκως διαφερόμενον ἑωυτῷ ὁμολογέει· παλίντροπος ἁρμονίη ὅκωσπερ τόξου καὶ λύρης.

XLVI. Aristoteles Eth. Nic. viii. 2, p. 1155 b 1. καὶ περὶ αὐτῶν τούτων ἀνώτερον ἐπιζητοῦσι καὶ φυσικώτερον· Εὐριπίδης μὲν φάσκων ἐρᾶν μὲν ὄμβρου γαῖαν ξηρανθεῖσαν, ἐρᾶν δὲ σεμνὸν οὐρανὸν πληρούμενον ὄμβρου πεσεῖν ἐς γαῖαν· καὶ Ἡράκλειτος τὸ ἀντίξουν συμφέρον, καὶ ἐκ τῶν διαφερόντων καλλίστην ἁρμονίαν, καὶ πάντα κατ' ἔριν γίνεσθαι.

XLVII. Ἁρμονίη ἀφανὴς φανερῆς κρείσσων.

XLVIII. Μὴ εἰκῆ περὶ τῶν μεγίστων συμβαλώμεθα.

XLIX. Χρὴ εὖ μάλα πολλῶν ἵστορας φιλοσόφους ἄνδρας εἶναι.

L. Γναφέων ὁδὸς εὐθεῖα καὶ σκολιὴ μία ἐστὶ καὶ ἡ αὐτή.

LI. Ὄνοι σύρματ' ἂν ἕλοιντο μᾶλλον ἢ χρυσόν.

LII. Θάλασσα ὕδωρ καθαρώτατον καὶ μιαρώτατον, ἰχθύσι μὲν πότιμον καὶ σωτήριον, ἀνθρώποις δὲ ἄποτον καὶ ὀλέθριον.

LIII. Columella de R. R. viii. 4 : siccus etiam pulvis et cinis, ubicunque cohortem porticus vel tectum protegit, iuxta parietes reponendus est, ut sit quo aves se perfundant: nam his rebus plumam pinnasque emendant, si modo credimus Ephesio Heraclito qui ait: sues coeno, cohortales aves pulvere (vel cinere) lavari.

LIV. Βορβόρῳ χαίρειν.

LV. Πᾶν ἑρπετὸν πληγῇ νέμεται.

LVI. Παλίντροπος ἁρμονίη κόσμου ὅκωσπερ λύρης καὶ τόξου.

LVII. Ἀγαθὸν καὶ κακὸν ταὐτόν.

LVIII. Hippolytus Ref. haer. ix. 10: καὶ ἀγαθὸν καὶ κακύν (scil. ἕν ἐστι)· οἱ γοῦν ἰατροί, φησὶν ὁ Ἡράκλειτος, τέμνοντες καίοντες πάντη βασανίζοντες κακῶς τοὺς ἀρρωστοῦντας ἐπαιτιῶνται μηδέν' ἄξιον μισθὸν λαμβάνειν παρὰ τῶν ἀρρωστούντων, ταῦτα ἐργαζόμενοι τὰ ἀγαθὰ καὶ † τὰς νόσους †.

LIX. Συνάψειας οὖλα καὶ οὐχὶ οὖλα, συμφερόμενον διαφερόμενον, συνᾷδον διᾷδον· ἐκ πάντων ἓν καὶ ἐξ ἑνὸς πάντα.

LX. Δίκης οὔνομα οὐκ ἂν ᾔδεσαν, εἰ ταῦτα μὴ ἦν.

LXI. Schol. B. in Il. iv. 4, p. 120 Bekk.: ἀπρεπές φασιν, εἰ τέρπει τοὺς θεοὺς πολέμων θέα. ἀλλ' οὐκ ἀπρεπές· τὰ γὰρ γενναῖα ἔργα τέρπει. ἄλλως τε πόλεμοι καὶ μάχαι ἡμῖν μὲν δεινὰ δοκεῖ, τῷ δὲ θεῷ οὐδὲ ταῦτα δεινά. συντελεῖ γὰρ ἅπαντα ὁ θεὸς πρὸς ἁρμονίαν τῶν ὅλων, οἰκονομῶν τὰ συμφέροντα, ὅπερ καὶ Ἡράκλειτος λέγει, ὡς τῷ μὲν θεῷ καλὰ πάντα καὶ ἀγαθὰ καὶ δίκαια, ἄνθρωποι δὲ ἃ μὲν ἄδικα ὑπειλήφασιν, ἃ δὲ δίκαια.

LXII. Εἰδέναι χρὴ τὸν πόλεμον ἐόντα ξυνόν, καὶ δίκην ἔριν· καὶ γινόμενα πάντα κατ' ἔριν καὶ † χρεώμενα †.

LXIII. Ἔστι γὰρ εἱμαρμένα πάντως * * * *.

LXIV. Θάνατός ἐστι ὁκόσα ἐγερθέντες ὁρέομεν, ὁκόσα δὲ εὕδοντες ὕπνος.

LXV. Ἓν τὸ σοφὸν μοῦνον· λέγεσθαι οὐκ ἐθέλει καὶ ἐθέλει Ζηνὸς οὔνομα.

LXVI. Τοῦ βιοῦ οὔνομα βίος, ἔργον δὲ θάνατος.

LXVII. Ἀθάνατοι θνητοί, θνητοὶ ἀθάνατοι, ζῶντες τὸν ἐκείνων θάνατον τὸν δὲ ἐκείνων βίον τεθνεῶτες.

LXVIII. Ψυχῇσι γὰρ θάνατος ὕδωρ γενέσθαι, ὕδατι δὲ θάνατος γῆν γενέσθαι· ἐκ γῆς δὲ ὕδωρ γίνεται, ἐξ ὕδατος δὲ ψυχή.

LXIX. Ὁδὸς ἄνω κάτω μία καὶ ὡυτή.

LXX. Ξυνὸν ἀρχὴ καὶ πέρας.

LXXI. Ψυχῆς πείρατα οὐκ ἂν ἐξεύροιο πᾶσαν ἐπιπορευόμενος ὁδόν.

LXXII. Ψυχῇσι τέρψις ὑγρῇσι γενέσθαι.

LXXIII. Ἀνὴρ ὁκότ' ἂν μεθυσθῇ, ἄγεται ὑπὸ παιδὸς ἀνήβου σφαλλόμενος, οὐκ ἐπαΐων ὅκη βαίνει, ὑγρὴν τὴν ψυχὴν ἔχων.

LXXIV. Αὔη ψυχὴ σοφωτάτη καὶ ἀρίστη.

LXXV. † Αὐγὴ ξηρὴ ψυχὴ σοφωτάτη καὶ ἀρίστη †.

LXXVI. † Οὗ γῆ ξηρή, ψυχὴ σοφωτάτη καὶ ἀρίστη †.

LXXVII. Ἄνθρωπος, ὅκως ἐν εὐφρόνῃ φάος, ἅπτεται ἀποσβέννυται.

LXXVIII. Plutarchus Consol. ad Apoll. 10, p. 106: πότε γὰρ ἐν ἡμῖν αὐτοῖς οὐκ ἔστιν ὁ θάνατος; καὶ ᾗ φησιν Ἡράκλειτος, ταῦτ' εἶναι ζῶν καὶ τεθνηκός, καὶ τὸ ἐγρηγορὸς καὶ τὸ καθεῦδον, καὶ νέον καὶ γηραιόν· τάδε γὰρ μεταπεσόντα ἐκεῖνά ἐστι κἀκεῖνα πάλιν μεταπεσόντα ταῦτα.

LXXIX. Αἰὼν παῖς ἐστι παίζων πεσσεύων· παιδὸς ἡ βασιληίη.

LXXX. Ἐδιζησάμην ἐμεωυτόν.

LXXXI. Ποταμοῖσι τοῖσι αὐτοῖσι ἐμβαίνομέν τε καὶ οὐκ ἐμβαίνομεν, εἰμέν τε καὶ οὐκ εἰμεν.

LXXXII. Κάματός ἐστι τοῖς αὐτοῖς μοχθεῖν καὶ ἄρχεσθαι.

LXXXIII. Μεταβάλλον ἀναπαύεται.

LXXXIV. Καὶ ὁ κυκεὼν διίσταται μὴ κινεόμενος.

LXXXV. Νέκυες κοπρίων ἐκβλητότεροι.

LXXXVI. Γενόμενοι ζώειν ἐθέλουσι μόρους τ' ἔχειν· μᾶλλον δὲ ἀναπαύεσθαι, καὶ παῖδας καταλείπουσι μόρους γενέσθαι.

LXXXVII. Plutarchus de Orac. def. 11, p. 415: οἱ μὲν "ἡβῶντος" ἀναγινώσκοντες (apud Hesiod. fr. 163 Goettling) ἔτη τριάκοντα ποιοῦσι τὴν γενεὰν καθ' Ἡράκλειτον· ἐν ᾧ χρόνῳ γεννῶντα παρέχει τὸν ἐξ αὑτοῦ γεγεννημένον ὁ γεννήσας.

LXXXVIII. Io. Lydus de Mensibus iii. 10, p. 37 ed. Bonn: ὁ τριάκοντα ἀριθμὸς φυσικώτατός ἐστιν· ὁ γὰρ ἐν μονάσι τριάς, τοῦτο ἐν δεκάσι τριακοντάς. ἐπεὶ καὶ ὁ τοῦ μηνὸς κύκλος συνέστηκεν ἐκ τεσσάρων τῶν ἀπὸ μονάδος ἑξῆς τετραγώνων α΄, δ΄, θ΄, ιϛ΄. ὅθεν οὐκ ἀπὸ σκοποῦ Ἡράκλειτος γενεὰν τὸν μῆνα καλεῖ.

LXXXIX. Ex homine in tricennio potest avus haberi.

XC. M. Antoninus vi. 42: πάντες εἰς ἓν ἀποτέλεσμα συνεργοῦμεν, οἱ μὲν εἰδότως καὶ παρακολουθητικῶς, οἱ δὲ ἀνεπιστάτως· ὥσπερ καὶ τοὺς καθεύδοντας, οἶμαι, ὁ Ἡράκλειτος ἐργάτας εἶναι λέγει καὶ συνεργοὺς τῶν ἐν τῷ κόσμῳ γινομένων.

XCI. Ξυνόν ἐστι πᾶσι τὸ φρονέειν. ξὺν νόῳ λέγοντας ἰσχυρίζεσθαι χρὴ τῷ ξυνῷ πάντων, ὅκωσπερ νόμῳ πόλις καὶ πολὺ ἰσχυροτέρως. τρέφονται γὰρ πάντες οἱ ἀνθρώπειοι νόμοι ὑπὸ ἑνὸς τοῦ θείου· κρατέει γὰρ τοσοῦτον ὁκόσον ἐθέλει καὶ ἐξαρκέει πᾶσι καὶ περιγίνεται.

XCII. Τοῦ λόγου δ' ἐόντος ξυνοῦ, ζώουσι οἱ πολλοὶ ὡς ἰδίην ἔχοντες φρόνησιν.

XCIII. Ὧι μάλιστα διηνεκέως ὁμιλέουσι, τούτῳ διαφέρονται.

XCIV. Οὐ δεῖ ὥσπερ καθεύδοντας ποιεῖν καὶ λέγειν.

XCV. Plutarchus de Superst. 3, p. 166: ὁ Ἡράκλειτός φησι, τοῖς ἐγρηγορόσιν ἕνα καὶ κοινὸν κόσμον εἶναι, τῶν δὲ κοιμωμένων ἕκαστον εἰς ἴδιον ἀποστρέφεσθαι.

XCVI. Ἦθος γὰρ ἀνθρώπειον μὲν οὐκ ἔχει γνώμας, θεῖον δὲ ἔχει.

XCVII. Ἀνὴρ νήπιος ἤκουσε πρὸς δαίμονος ὅκωσπερ παῖς πρὸς ἀνδρός.

XCVIII. Plato Hipp. mai. 289 B: ἢ οὐ καὶ Ἡράκλειτος ταὐτὸν τοῦτο λέγει, ὃν σὺ ἐπάγει, ὅτι ἀνθρώπων ὁ σοφώτατος πρὸς θεὸν πίθηκος φανεῖται καὶ σοφίᾳ καὶ κάλλει καὶ τοῖς ἄλλοις πᾶσιν;

XCIX. Plato Hipp. mai. 289 A: ὦ ἄνθρωπε, ἀγνοεῖς ὅτι τὸ τοῦ Ἡρακλείτου εὖ ἔχει, ὡς ἄρα πιθήκων ὁ κάλλιστος αἰσχρὸς ἄλλῳ γένει συμβάλλειν, καὶ χυτρῶν ἡ καλλίστη αἰσχρὰ παρθένων γένει συμβάλλειν, ὥς φησιν Ἱππίας ὁ σοφός.

C. Μάχεσθαι χρὴ τὸν δῆμον ὑπὲρ τοῦ νόμου ὅκως ὑπὲρ τείχεος.

CI. Μόροι γὰρ μέζονες μέζονας μοίρας λαγχάνουσι.

CII. Ἀρηιφάτους θεοὶ τιμῶσι καὶ ἄνθρωποι.

CIII. Ὕβριν χρὴ σβεννύειν μᾶλλον ἢ πυρκαϊήν.

CIV. Ἀνθρώποισι γίνεσθαι ὁκόσα θέλουσι οὐκ ἄμεινον. νοῦσος ὑγίειαν ἐποίησε ἡδὺ καὶ ἀγαθόν, λιμὸς κόρον, κάματος ἀνάπαυσιν.

CV. Θυμῷ μάχεσθαι χαλεπόν· ὅ τι γὰρ ἂν χρηίζῃ γίνεσθαι, ψυχῆς ὠνέεται.

CVI. †Ἀνθρώποισι πᾶσι μέτεστι γιγνώσκειν ἑαυτοὺς καὶ σωφρονεῖν†.

CVII. †Σωφρονεῖν ἀρετὴ μεγίστη· καὶ σοφίη ἀληθέα λέγειν καὶ ποιεῖν κατὰ φύσιν ἐπαίοντας†.

CVIII. Ἀμαθίην ἄμεινον κρύπτειν· ἔργον δὲ ἐν ἀνέσει καὶ παρ' οἶνον.

CIX. †Κρύπτειν ἀμαθίην κρέσσον ἢ ἐς τὸ μέσον φέρειν†.

CX. Νόμος καὶ βουλῇ πείθεσθαι ἑνός.

CXI. Τίς γὰρ αὐτῶν νόος ἢ φρήν; [δήμων] ἀοιδοῖσι ἕπονται καὶ διδασκάλῳ χρέωνται ὁμίλῳ, οὐκ εἰδότες ὅτι πολλοὶ κακοὶ ὀλίγοι δὲ ἀγαθοί. αἱρεῦνται γὰρ ἓν ἀντία πάντων οἱ ἄριστοι, κλέος ἀέναον θνητῶν, οἱ δὲ πολλοὶ κεκόρηνται ὅκωσπερ κτήνεα.

CXII. Ἐν Πριήνῃ Βίας ἐγένετο ὁ Τευτάμεω, οὗ πλέων λόγος ἢ τῶν ἄλλων.

CXIII. Εἷς ἐμοὶ μύριοι, ἐὰν ἄριστος ᾖ.

CXIV. Ἄξιον Ἐφεσίοις ἡβηδὸν ἀπάγξασθαι πᾶσι καὶ τοῖς ἀνήβοις τὴν πόλιν καταλιπεῖν, οἵτινες Ἑρμόδωρον ἄνδρα ἑωυτῶν ὀνήιστον ἐξέβαλον, φάντες · ἡμέων μηδὲ εἷς ὀνήιστος ἔστω, εἰ δὲ μή, ἄλλῃ τε καὶ μετ' ἄλλων.

CXV. Κύνες καὶ βαΰζουσι ὃν ἂν μὴ γινώσκωσι.

CXVI. Ἀπιστίη διαφυγγάνει μὴ γινώσκεσθαι.

CXVII. Βλὰξ ἄνθρωπος ἐπὶ παντὶ λόγῳ ἐπτοῆσθαι φιλέει.

CXVIII. Δοκεόντων ὁ δοκιμώτατος γινώσκει πλάσσειν · καὶ μέντοι καὶ δίκη καταλήψεται ψευδέων τέκτονας καὶ μάρτυρας.

CXIX. Diogenes Laert. ix. 1: τόν θ' Ὅμηρον ἔφασκεν ἄξιον ἐκ τῶν ἀγώνων ἐκβάλλεσθαι καὶ ῥαπίζεσθαι, καὶ Ἀρχίλοχον ὁμοίως.

CXX. Unus dies par omni est.

CXXI. Ἦθος ἀνθρώπῳ δαίμων.

CXXII. Ἀνθρώπους μένει τελευτήσαντας ἄσσα οὐκ ἔλπονται οὐδὲ δοκέουσι.

CXXIII. Ἔνθαδε ἐόντας ἐπανίστασθαι καὶ φύλακας γίνεσθαι ἐγερτὶ ζώντων καὶ νεκρῶν.

CXXIV. Νυκτιπόλοι, μάγοι, βάκχοι, λῆναι, μύσται.

CXXV. Τὰ γὰρ νομιζόμενα κατ' ἀνθρώπους μυστήρια ἀνιερωστὶ μυεῦνται.

CXXVI. Καὶ τοῖς ἀγάλμασι τουτέοισι εὔχονται, ὁκοῖον εἴ τις τοῖς δόμοισι λεσχηνεύοιτο, οὔ τι γινώσκων θεοὺς οὐδ' ἥρωας, οἵτινές εἰσι.

CXXVII. Εἰ μὴ γὰρ Διονύσῳ πομπὴν ἐποιεῦντο καὶ ὕμνεον ᾆσμα αἰδοίοισι, ἀναιδέστατα εἴργαστ' ἄν · ὡυτὸς δὲ Ἀίδης καὶ Διόνυσος, ὅτεῳ μαίνονται καὶ ληναΐζουσι.

CXXVIII. Iamblichus de Myst. v. 15: θυσιῶν τοίνυν τίθημι διττὰ εἴδη · τὰ μὲν τῶν ἀποκεκαθαρμένων παντάπασιν ἀνθρώπων, οἷα ἐφ' ἑνὸς ἄν ποτε γένοιτο σπανίως, ὥς φησιν Ἡράκλειτος, ἤ τινων ὀλίγων · εὐαριθμήτων ἀνδρῶν · τὰ δ' ἔνυλα καὶ σωματοειδῆ καὶ διὰ μεταβολῆς συνιστάμενα, οἷα τοῖς ἔτι κατεχομένοις ὑπὸ τοῦ σώματος ἁρμόζει.

CXXIX. Ἄκεα.

CXXX. Καθαίρονται δὲ αἵματι μιαινόμενοι ὥσπερ ἂν εἴ τις ἐς πηλὸν ἐμβὰς πηλῷ ἀπονίροιτο.

RD 108

Deacidified using the Bookkeeper process
Neutralizing agent: Magnesium Oxide
Treatment Date: July 2004

PreservationTechnologies
A WORLD LEADER IN PAPER PRESERVATION
111 Thomson Park Drive
Cranberry Township, PA 16066
(724) 779-2111